P9-EML-865

Work Meets Life

Work Meets Life

Exploring the Integrative Study of Work in Living Systems

edited by Robert Levin, Simon Laughlin, Christina De La Rocha, and Alan Blackwell

The MIT Press
Cambridge, Massachusetts
London, England

RECEIVED

JAN 3 1 2012

MINNESOTA STATE UNIVERSITY LIBRARY
MANKATO, MN 56002-8419

© 2011 Massachusetts Institute of Technology

All rights reserved. No part of this book may be reproduced in any form by any electronic or mechanical means (including photocopying, recording, or information storage and retrieval) without permission in writing from the publisher.

MIT Press books may be purchased at special quantity discounts for business or sales promotional use. For information, please email special_sales@mitpress.mit.edu or write to Special Sales Department, The MIT Press, 55 Hayward Street, Cambridge, MA 02142.

This book was set in Stone Sans and Stone Serif by Toppan Best-set Premedia Limited. Printed and bound in the United States of America.

Library of Congress Cataloging-in-Publication Data

Work meets life : exploring the integrative study of work in living systems / edited by Robert Levin ... [et al.].
   p. cm.
Includes bibliographical references and index.
ISBN 978-0-262-01412-0 (hardcover : alk. paper)
1. Biological systems. 2. Work. 3. Bioenergetics. I. Levin, Robert.
QH313.W67 2011
571—dc22
                                                                    2009037133

10 9 8 7 6 5 4 3 2 1

QH
313
.W67
2011

7459109

To those who showed us how to work,

To those who taught us of their work,

To those who helped us by their work.

Contents

Preface

My coeditors have asked me to set out a few key aspects of how this volume came into being. Doing so also gives me the opportunity to properly acknowledge their contributions to this volume.

My own earliest interest in the approach developed in this volume began over twenty-five years ago, during a literature search on the physiology and cellular biochemistry of human hypothermia. By the kind of happy coincidence more likely when reviewing the printed *Biological Abstracts* of the day, I found together, under the heading for hypothermia, both an article on the effect of cold temperature on a critical energetic membrane pump (described in chapter 1) and an article on the epidemiology of human hypothermia. From the second article, I learned that hypothermia most frequently afflicts not mountaineers but individuals in urban settings, including those homeless and unemployed sleeping under bridges.

It seemed to me at the time that those two articles represented two widely separated, but related, aspects of the domain of work, in natural and human life. The research on cold effects on membrane pumps represented an insight about work from basic biological research and demonstrated that effects of work at the molecular level can have profound impact on the whole organism. The epidemiological research indicated that practical problems in work and employment can directly affect individuals' lives and their survival at a biological level.

Ever since that time, I have been trying to pull such apparently disparate areas of the domain of work closer together. That task has been impossible to accomplish as an individual, but it is one that the group of editors and contributors for this volume has been able to begin.

That group has come together over a period of twenty years in a variety of interdisciplinary settings, through:

• Work with Anne Bekoff, Joe Rosse, and Jim Williams beginning in 1989, and later with Stacy Saturay, through the University of Colorado at Boulder (CU-Boulder) Research Park.

• Various interdisciplinary symposia at CU-Boulder at which Norm Pace, Steve Maier, and Mike Lightner were speakers.
• The proposal, approval, and founding of the Center for the Integrative Study of Work (CISW) at CU-Boulder in 2000, which built on the work conducted earlier through the CU-Boulder Research Park.
• Work with Kevin Laland and Simon Laughlin, beginning in 2001, at the Department of Zoology and Subdepartment of Animal Behavior at the University of Cambridge.
• Interdisciplinary symposia at the University of Cambridge with Christina De La Rocha, Alan Blackwell, and Gillian Brown.

These in turn led to the development of two working conferences held in 2003 and 2004, as well as to the development of a symposium titled "Putting Energy and Information to Work in Living Systems" at the 2003 American Association for the Advancement of Science annual meeting in Denver. Together, these formed the initial foundation for this volume. The first of the two working conferences was held in Boulder in the summer of 2003 and the second in Cambridge in the summer of 2004. The working conferences followed a format proposed by Willy Brown, master of Darwin College, Cambridge. We did not present papers at either working conference. At the first working conference, in Boulder, we each discussed our research areas and the ways these areas might relate to the study of work. We also hashed out some questions about a foundation for a study of work. The stimulating research presentations and the vigorous enthusiasm of each participant for the importance of the research that had been presented by other participants shaped the flavor of the volume, and it also put the kibosh on preconceptions (my own) for a more conceptual foundation for the volume. At the second conference, in Cambridge in 2004, each of our partially drafted chapters was the subject of intense discussion sessions. Some participants were not volume contributors, so we were fortunate to get early input of how those outside of the project might view this effort.

I met Bob Prior of the MIT Press at a conference in 2004. He asked me what I was working on, and I told him about *Work Meets Life*. Very characteristically of Bob, as we later came to know him, he soon offered insightful and helpful suggestions about the book and about the possibilities for the integrative approach we were pursuing to our exploration of work. It was clear that Bob was not trying to force the volume into a preexisting mold. It also was clear that such an approach would be vital to developing the volume successfully, and we were all therefore pleased to bring the project to Bob and to the MIT Press. Many of Bob's recommendations find form in the volume, and so do his solutions for every difficulty that arose.

As I write this, in 2010, the contributors and the editors have been revising and shaping the contributions into a volume that truly integrates the study of work in living systems across many domains. The efforts of the individual contributors in this way have been remarkable: They have developed chapters that are each integrative in their own right—a development that went beyond what we anticipated in our early planning.

My coeditors contributed greatly to these developments, and it is a privilege to have the opportunity to work with them. Simon Laughlin, Christina De La Rocha, and Alan Blackwell gave of their time, of their expertise in their own areas of research and knowledge, and of their interest in and skill at developing well-written and well-thought-out chapters. They each contributed considerable talent and experience at exploring new scientific and intellectual regions, for which they are each known in their research areas. As well, because this volume required that they work as editors on a very wide range of subject areas, they always approached contributions and contributors from outside of their own areas with great respect. But never with deference, because each recognized that they were forerunners for the readers of this volume.

Simon Laughlin helped to plan the first conference around that approach and as an active participant throughout that conference emphasized the importance of putting the volume on the solid footing of research-based fundamentals. Since that time, he has provided insights to his editing and his development of material from his extensive knowledge of energy and information in many areas and from his intensely developed knowledge of a range of other subject areas. Simon edited the Reflections at the end of this volume in a way that brings a wide range of knowledge and perspectives together into one body of work.

Christina De La Rocha participated in and helped to plan the first conference and then organized and hosted the second conference in Cambridge. Once manuscript development was underway, she was an active editor, working with successive versions of manuscripts for every chapter in this book. She was often the first editor to provide a review of a submitted manuscript, and she was always willing to read one more draft for a contributor. In addition, from the inception of the project, Christina worked to pull together material under discussion to be useful to our readers later on. Together with her own thoughts and analysis, she used that material to develop a unique editorial perspective on how work gets done in living systems, which appears as the Observations portion of the Reflections section.

Alan Blackwell became an editor through his continual interest and work on the project that went well beyond his role as a contributor. He drew on his unique perspective as a psychologist, engineer, and computer scientist to comment insightfully on manuscripts from a range of areas. In addition, Alan contributed his considerable skills as a designer and as an illustrator; his contributions of this kind appear in the volume in several places—from the outset to the very end. Every one of Alan's illustrations was developed from discussions similar to a design interview, and out of these discussions emerged not only illustrations but new perspectives and content for the book. Alan used this approach in developing the Design Interlude portion of the Reflections section as an editorial contribution. The illustration that forms the core of the Design Interlude is a remarkable one-page visual representation of concepts developed throughout this volume.

The contributors to this volume and I join together in thanking Simon, Christina, and Alan for all of their contributions, which reach well beyond what I have mentioned here. Likewise, thanks from Simon, Christina, Alan, and me go to the contributors, who each went well beyond what we expected. My own personal thanks to the editors and to the contributors: I learned a great deal from working on each of the chapters, and doing so has been a privilege. I learned the most from having the opportunity to work with each of the editors and contributors. If, in turn, we have been able to transmit some portion of what we have all learned from this endeavor through this volume, then that will be our measure of success.

Robert Levin

Acknowledgments

Many individuals contributed to the progress and completion of this volume. Three individuals who are not named contributors to this volume played central roles in its earliest formulation and development and have continued to contribute to this project over a period of many years.

James F. Williams, II laid the foundation for the network of relationships that made this volume possible, contributed to its content and to the development of a process for developing interdisciplinary research teams, and developed the formulation that work could be studied in many different systems by asking the question: How does work get done?

Anne Bekoff provided the first discussions of the approach and content that underlie this volume. In those discussions, she recommended the formulation for an integrative approach, based on but not limited to integrative biology, for studying work in living systems. Over many years, she has helped to apply this approach to the many steps involved in developing this volume.

Norman Pace made a unique contribution to our first conference that became a central concept for the volume: the idea that work could be seen as a fundamentally integrative process, starting with the work of chemiosmosis in the cell, rather than creating arguments by analogy between different forms of work. Afterward, Norm continued to advise and participate in the development of this foundation and its contribution to the volume.

Four of the contributing authors, Kevin Laland, Michael Lightner, Steven Maier, and Joseph Rosse, made substantive editorial contributions to this volume.

The volume has also benefited from the editorial contributions of Bridget Julian, Kathy Kaiser, Alice Levine, and Faan Tone Liu, each of whose contributions were supremely timely and helpful.

Our volume developed in and around two working group conferences, and we are grateful for the contributions of the following participants: Marie Banich, Anne Bekoff, David Grant, John Hoffecker, John Odling-Smee, Jennifer Whyte, and Kenneth Wright.

The first of the two conferences from which this book developed, held at CU-Boulder, was made possible by the cooperation of the Institute of Cognitive Science, which provided the facilities, and by the efforts of Donna Caccamise, the associate director of the Institute of Cognitive Science, and Jean Bowen, the conference organizer. The conference was funded in part by WorkScience and in part by the Center for the Integrative Study of Work (CISW) at CU-Boulder.

The second conference, at the University of Cambridge, was made possible by the cooperation of the Department of Earth Sciences, which provided the facilities and partial funding, together with partial funding by WorkScience. Facilities were also provided by Darwin College, Cambridge.

The initial collaborative work that preceded the volume itself, as well as work on the volume, was made possible by facilities generously provided to the senior editor by the Department of Zoology at the University of Cambridge, including the Subdepartment of Animal Behavior, Madingley, and the Museum of Zoology, and by Darwin College, Cambridge. Facilities for development of this volume have also been provided over the span of the project by Engineering Research East, CU-Boulder. Facilities and professional assistance for conducting library research for this volume have been provided throughout this project at Norlin Library, CU-Boulder. Members of the CINC group at CU-Boulder helped make it possible to successfully tackle many challenging aspects of completing this volume. The members of the Center for the Integrative Study of Work at CU-Boulder always provided an intellectually stimulating environment for discussing this project throughout.

We acknowledge the intellectual contributions and the help and support of the individuals named above and of: Michael Akam, Stephen Arch, Immanuel Barshi, Dennis Bray, Nate Brown, Willy Brown, Joseph Falke, Kathleen Farrell, Adrian Friday, Todd Gleeson, Larry Gold, Leo Howe, Sheryl Jensen, Rufus Johnstone, Judah Levine, Clayton Lewis, Diane McKnight, Karen Meyer-Arendt, Christine Mueller, Jeremy Niven, Lynn Parisi, Jerry Peterson, James Reuler, Laurent Seugnet, Marla Shapiro, Neil Tickner, Linda Watkins, Ted Weverka, and Linda Wheatley.

We close by thanking three groups of people:

First, the editors and professionals at the MIT Press. In particular, we thank Executive Editor Robert V. Prior—whose contributions are described in the Preface. We recognize with appreciation the help and professionalism of Susan Buckley. We thank Mel Goldsipe and Mary Reilly, who worked to ensure that the quality of the book produced exceeded the quality of the manuscript received. We thank Tobiah Waldron, whose index conveys the integrative nature of this volume.

Second, to those who helped us and for any number of reasons are not named here—your help has been invaluable.

Finally, to those family and friends who gave us help and support during this process—you are close to us in our hearts.

Robert Levin
Simon Laughlin
Christina De La Rocha
Alan Blackwell

Introduction: A Fresh Perspective on Work

"The question is not *what is the answer*—the question is *what is the question.*"

This was a favorite and emphatic saying of one of our fathers—a dictum by which he steered his own professional work and by which he engaged with those with whom he worked. Those of us who have worked together and engaged with each other in developing this volume have done so not because we believe that the study of work in various areas is in need of new answers, but instead because we believe it is in need of new questions—new questions that can be asked now because of advances in knowledge in a range of scientific disciplines, including many advances that until now have not often been included in developing an understanding of how work gets done.

Our exploration of an integrative approach to work in living systems is therefore an invitation to explore new country and thereby find new perspectives and new vistas. Our volume is intended as a guidebook, in the sense of providing opportunities to find and explore these new perspectives. Our volume is not a handbook, because it is neither authoritative nor is it comprehensive—nor is that its intent. This volume is instead the product of a group of people exploring territory based on their own individual interests, and some mutually discovered interests, in order to help create new questions about work.

We hope that readers will find new questions through exploring this book and through their own explorations and that, by sharing these new questions together, an integrative approach to studying work in living systems can develop, and from which new answers may someday come. We believe that continuing to ask the question— *How does work get done?*—in a variety of settings and systems can lay the foundation for an integrative approach to studying work in living systems, including human beings.

The purpose of this book, then, is to explore the possibility of an integrative study of work in living systems, one that more broadly incorporates and utilizes ongoing developments in a range of basic scientific and research disciplines, some of which are described in this volume. In this volume, you will find a collection of chapters from researchers of widely varied backgrounds, including different fields in biological and earth sciences, psychological and social sciences, and engineering. These researchers have taken areas from their own research to provide perspectives on how work gets done in living systems. The chapters here, for example, are drawn from such areas as cellular bioenergetics, sensory neurobiology, electrical and computer engineering, biogeochemistry, evolutionary biology, design and cognitive science, industrial-organizational psychology, and an integrative branch of neuroscience—psychoneuroimmunology. Each of the chapters is intended to illuminate some aspect of how work gets done in living systems. Because of the variety of research represented, we believe that you will be rewarded throughout the book by being able to explore perspectives on work, and on living systems, that you might not have considered before. That certainly has been the experience of each of us who have participated in developing this volume.

During the two working conferences held at the outset of this exploration, the participants grappled with the question of how to present research from a wide range of disciplines to the range of academic readers who may come to this volume. We decided that this could be done most effectively by each contributor carving a "cameo" of research that the contributor felt would make an important contribution to the study of work—and that could connect with research in other areas. Our chapters take this form.

Our hope is that these cameos—these carved stones—have been so shaped by the authors and placed by the editors such that the resulting volume results in an image that had not been evident before, of different aspects of work as it is performed in living systems.

That is, again, why we have styled the book as a guidebook—a book from which you can pick out interesting perspectives that can help you in your own explorations, rather than as a handbook styled to authoritatively explain what should be interesting and worth pursuing. The most important connections to us are therefore not so much the ones that we have seen as the ones that you may make as you read the book. We hope this book provides you with some new knowledge and, more importantly, with new perspectives. The new domains we hope you will come to explore through this book will be the connections that you make between the wonder of work and the wonder of life.

The contributors have taken particular care in shaping their chapters to the purpose of the book. Therefore, as much as possible from this point forward, we want to simply invite readers to explore. The beginning of the book includes perspectives on some fundamentals of work in living systems. Toward the end of the book, in the Reflections section, we report on some things that the authors have taken away from this exploration and on questions the authors would like to pose. In between, we hope you will find facets of work and facets of life that will at times gleam in a new way, as you find and develop connections of how work meets life.

Exploring an Integrative Study of Work in Living Systems

Something we thought we owed readers setting off on this exploration was what we might mean by *the integrative study of work in living systems.*

Integrative

By *integrative*, we mean the word in the sense the term is used in biology, rather than in a merely holistic sense. An integrative study in biology looks at the same phenomenon at different levels of analysis and of biological organization, and then seeks to identify connections between these levels—without ever assuming that connections must exist. For instance, one might look at the effect of temperature stress at the biochemical, cellular, and organismic levels. After the effects have been explored at these individual levels, then one can look for possible connections between the levels. But one never assumes a priori that connections exist. In this volume, the explorations at different levels of organization are made by each author. The connections emerge in relating one chapter to another. Taking an integrative approach does not assume that all aspects of work in living systems are connected to one another.

Study

We have used the word integrative *study* of work rather than integrative *science* of work for three reasons. First, it is a reminder that the work in this volume is deliberately exploratory rather than definitive. Second, we have included a wide range of disciplines in this volume that do involve rigorous study but are not necessarily "science" disciplines in the traditional sense. Third, we thought that some humility might be in order before branding this approach as a science.

Work

The word *work* may be the word that one might most expect to find explicated definitively at the outset of this volume. Deliberately, that is not the case. *Work* turns out to be a word that in English has a very robust meaning over time. In the *Oxford English Dictionary*, for example, all of the meanings of *work*, presented in numerous definitions, are related in one way or another to the basic concept of exertion by which one gains sustenance—a very biologically connected meaning—or to some other closely related meaning. This is perhaps the more remarkable given that the word *work* has over 70 main definitions (as a noun and as a verb) in the *Oxford English Dictionary* and that these various definitions span the history of the English language from the present day back in time until the word *werk* disappears into Old High German and Latin. It is truly a rich concept, and it is a robust one, for the meanings revolve in one way or another around this same central concept.

We have left to those chapter authors who wanted to define *work* in their particular setting the opportunity to do so. We realize that some readers might prefer a central definition. We felt that providing an a priori definition at the outset of our exploration—either the actual exploration that the authors have now completed or the account of the explorations that this book provides—would be both presumptuous and limiting. We also chose to follow the approach of many astrobiology researchers who do not define *life*. Instead, exploring phenomena in the absence of a specific definition aids in their ability to characterize and understand a widely ranging phenomenon. Some authors in this volume have integrated their interests in defining work into their chapters. (For example, see Kevin Laland and Gillian Brown's approach in chapter 5.) The Reflections section at the end of the book provides readers with a sense of what kinds of enriched answers to the question of describing work might emerge from the kind of exploration of work that we have started.

Living Systems

Our choice of the term *living systems* to describe the scope of work has to do with the emphasis of this book and some of the connections we hoped to find and develop. Since we are writing about a phenomenon as rich as work, we wanted to see how the tremendous diversity of work on Earth, in both humans and other living systems, could be related to some fundamentals of how work gets done in living organisms.

This emphasis caused us to look at aspects of work that might be common across the range of living systems—living systems that display a wide range of strategies that allow them to work in this world. For instance, we have identified the biological foundation created by the energetic process known as *chemiosmosis* as one key process (to which we will turn in chapter 1). We have introduced the biological fundamental of a trade-off between energy and information in a biological system, which shapes everything we can do, even in an information-technology-based workplace. And we have included an exploration of the effect of organisms' work and activity on evolution, a process known as niche construction.

This volume therefore endeavors to introduce at least one new question, one that can be asked of many different organisms, including humans, at many different levels of organization, and from the starting points of different disciplines in disparate fields: *How does work get done?* The individual contributions to this volume show how this question can be fruitfully asked and elaborated in many different forms in the context of research topics in a wide range of fields. While some of the cameos might seem, at first consideration, far removed from the study of work, we hope that by volume's end our readers will place these stones, as we do, as integral to a fuller and more complete understanding of the mosaic of how work meets life.

Chapter Summaries

The contributors to this book were originally assembled around two geographical locations—the University of Colorado at Boulder in the United States and the University of Cambridge in the United Kingdom—where the integrative approach to the study of work has been developing. All of the contributors have a connection to one of these two centers in some form. Contributors were initially invited by a possible connection between their own research and the development of an integrative approach to the study of work in living systems. All were invited to participate on the basis of their own research and of their interest in contributing to what an integrative approach would look like and could come to be. At no time did we invite someone to join the group simply to fill a gap in the subject matter. The content in the book, and its fit, is therefore that of a mosaic, rather than of a blueprint.

The development of the contributor group, through two working conferences and the development of the chapters, has led to an interweaving of the content throughout the chapters that follow, across levels of analysis and across disciplinary lines. Our volume reflects this interweaving in two ways unexpected at the outset. We initially expected that the integrative nature of this volume would occur after the chapters were contributed. However, the contributors outdid this initial conception. Each chapter is integrative in its own right, developing its subject across levels of organization and across domains of knowledge. We also expected that interconnections between the chapters would likewise be made by reviewing the completed chapters. Instead, as the contributors came to know each other and each other's work over the course of the project, they frequently noted connections between their own contributions and those in other chapters. Those have grown into an extensive series of cross-references noted in the chapters. In addition, this volume began its life conceptually as a multiauthored book; some of the chapters have been developed in ways that are consistent with that earlier concept.

The process of developing the foundation for an integrative approach to understanding work in living systems begins in chapter 1, by Robert Levin, which explores the most fundamental work process in living systems. The process known as chemiosmosis is a process of physical work that takes place at the core of our cells hundreds of thousands of times each second. It creates the energy "currency" that enables our cells to function and carry out complex processes, and it largely uses very simple molecules to do so. Moreover, we can trace the molecular heritage of this process and learn that it is an ancient energetic work process, one that has been enabling life to do work for as long as life has been in a form we can recognize. Tracing the molecular heritage also shows us that it is a very widespread process—so widespread that it is used by all forms of life as we know it. Understanding these energetic processes is therefore crucial to understanding work in living systems. More than that, these energetic processes underlie every other work process—the characteristics of chemios-mosis, in turn, shape the characteristics of other work. In this way, energetic processes become the first of several *performance envelopes* bounding the performance of even complex human work, which are developed throughout the book.

In chapter 2, Simon Laughlin brings together concepts that may appear discon-nected in modern postindustrial human work, but that he shows to be inseparable in the work of a biological system. The first element in the connection is energy, which is directly tied to the thermodynamic concept of work—energy is the capacity to perform work. It is also, as shown in chapter 1, part and parcel of biological work processes. The second element is information, something that we have commonly come to think of as quite distinct from energy. However, Laughlin has utilized the photoreceptors of the compound eyes of three species of flies as a unique way to study and answer a crucial question about energy and information in living systems: How much energy does it take for a system to receive, transmit, and process information? This is a significant question, for no living system can perform work without informa-tion to direct it. By utilizing a single photoreceptor in a fly, Laughlin can measure the energy expended when one solitary photoreceptor receives one *bit*—like a computer bit—of information from a small change in light level. Laughlin's results indicate that the amount of energy is not small—but instead similar to that used by human muscles under strenuous exercise. Whether at the level of a fly eye, or scaled to the level of a rat brain, or a human brain, the amount of energy required creates an inherent energy–information trade-off. More information requires disproportionately more energy; energetic constraints therefore create informational constraints. Modern humans are not exempt, at either the level of the individual or of society. Humans have brilliantly off-loaded our own informational requirements onto computers at tremendous ener-getic expense to create, maintain, and utilize the machines. Moreover, the informa-tional tasks we are now expected to perform by the design of computers often exceed our own information constraints, themselves a direct product of energy–information trade-offs.

In chapter 3, Michael Lightner shows us how inherent trade-offs more generally affect the performance of work. In a complement to Laughlin's approach in chapter 2, Lightner's work starts with trade-offs that occur in the design and manufacture of computer chips and then extends the finding to living systems. Lightner's findings about the implications of a design optimization method for chip manufacture called *design centering* focus on what most affects the performance of a designed and manu-factured system. Lightner found that attempting to increase the performance of any such system, for instance by improving its speed, would actually and inherently increase the failure rate of the chip. The solution is to always consider *yield* as a param-eter that is always present alongside any performance parameters. When one does that, one finds that there is an inherent performance–yield trade-off. Increasing performance decreases yield, and increasing yield can only be achieved by *decreasing* performance—and decreasing performance is something better done in a planned than an unplanned way. Shifting the approach to design to maximize yield instead of performance creates the approach known as design centering. There are significant implications for work processes in living systems as well. First, because the performance–yield trade-off must occur in any work process, these trade-offs occur in human and biological work processes—though they may have been invisible, ignored, or attributed to other causes. Second, performance–yield trade-offs and processes analogous to design centering likely occur in how living systems perform work, as Lightner shows in an extended analysis applying these methods to the work of the Pony Express and other "equine express" systems, providing a new take on classic optimality theory for biological systems.

In chapter 4, Christina De La Rocha looks at photosynthesis as a work process, one nearly as widespread and as ancient as chemiosmosis and with a tremendous impact on all ecosystems on earth. Photosynthesis exhibits a constraint at one crucial part of the process that, similarly to processes described in chapter 3, is a trade-off between performance and yield. The constraint in photosynthesis is one that has become more apparent as environmental conditions have changed: The process of photosynthesis is "stuck" on an evolutionary basis with certain characteristics, particularly those of an enzyme called "Rubisco," whose work processes were highly efficient in a global environment that was high in carbon dioxide. The ongoing changes in the carbon environment, in part created by photosynthesis, have created a situation in which Rubisco is increasingly inefficient. De La Rocha shows how, like industrial engineers, we can identify the inefficiency in the work process. Unlike an industrial engineer, the photosynthesizing organisms have not been able to implement a more efficient Rubisco. She also shows us the impact that the work of photosynthesis has on the work of the biosphere more generally. This impact of the work of organisms on their environment also provides a specific example—one with widespread impact—of an evolutionary process related to the work of organisms presented in the chapter that follows.

In chapter 5, Kevin Laland and Gillian Brown consider the relationship between evolution and work in a different way. The earlier chapters each deal with some evolutionary implications of work processes in living systems. Laland and Brown ask what effects organisms' work has on the process of evolution itself. Through the evolutionary process known as *niche construction*, Laland and Brown show how the work that organisms do in their environment—for instance by constructing shelter, by gathering food, and by moving through space—shapes both the organisms' environments and the organisms' niches themselves; in fact, both their own niches and the niches of other organisms. The work that organisms do has a profound effect on the process of evolution—as De La Rocha has shown with respect to the work of photosynthetic organisms in chapter 4. Laland and Brown continue beyond niche construction to show how research on human behavioral ecology can effectively be applied to human work, through new developments in energetic definitions of fitness that they discuss, which they connect with the "working energy/take-home energy trade–off hypothesis" presented in chapter 8.

In chapter 6, Alan Blackwell brings the theme of constructing niches and environments into the present-day work environment through his investigation of the work of designers. Some other organisms may work with tools, as Laland and Brown describe in chapter 5, and many other organisms extensively shape their environment. Blackwell's research explores what constitutes the modern work of design and provides an engaging approach for investigating modern human work in a way that the investigation of the work of design in chapter 6 can be an example. Blackwell explores the work of design from the perspective of research with designers from across the spectrum of modern work. He uses this unique perspective to show us the similarities of how designers accomplish the work of design in many different domains. In so doing, we come to understand how design is work, and how it is a kind of work accomplished both with the head and with the hands. Blackwell closes his chapter by investigating the ways that designers—and all humans—make choices or investments in allocating their attention to different work processes, attention investments that he indicates are trade-offs similar to those described by Laughlin and by Lightner.

In chapter 7, Joseph Rosse and Stacy Saturay, in their investigation into the effects of job satisfaction (or, more especially, job dissatisfaction) on work, point out the potential for performance, health, and well-being to be lost under less than psychologically optimal working conditions. In particular, Rosse and Saturay consider that the way workers cope with dissatisfaction has effects on not just the quality of their work but on their health and family life. The various types of responses that workers exhibit to dissatisfaction include measures ranging from the positive and proactive (e.g., actively trying to resolve the problems that are the source of the dissatisfaction) to the negative that seemingly lack benefit (e.g., retaliation, capitulation, and disengagement). Rosse and Saturay explore whether these behaviors might indeed be adaptive and why workers might engage in them even though they do not seem beneficial. Rosse and Saturay then turn their attention to the bidirectionality of stress in work–family conflict and consider the impacts of stress and dissatisfaction in one area spilling over into the other area and vice versa. They note that with the rise of the dual-income family (perhaps one with double the possibility of work–life conflict) has come a shift in workers' focus away from career orientation and toward maintaining an equitable work–life balance, a behavior that could itself be viewed as an adaptive mechanism for dealing with dissatisfaction by minimizing the chances of it occurring.

Chapter 8, an introduction to the working energy/take-home energy trade-off hypothesis by Robert Levin, Kevin Laland, and Stacy Saturay, originally developed in parallel with this book. The working energy/take-home energy trade-off hypothesis is based on a kind of hypothesis frequently used in biological research, especially in evolutionary biology: Researchers formulate a hypothesis that might be applied to a broad range of situations or organisms, that might then be tested against the hypothesis. This allows organisms to be compared with each other and with the hypothesis, leading to a better understanding of a given phenomenon. The working energy/take-home energy trade-off hypothesis begins from the foundation of energetic trade-offs by now well established in earlier chapters and explores what happens if the effects of those trade-offs are extended to one aspect of human work. Thus, the starting point for the hypothesis is that organisms respond to changes in the balance between two energetic requirements: the requirement for energy to fuel an organism's immediate needs (labeled as *working energy* in the chapter) and the requirement for energy that needs to be conserved for such uses as safety, reproduction, and providing for changes in the environment (labeled as *take-home energy*). All organisms have mechanisms allocating energy to meet these demands. One of the important implications of the working energy/take-home energy trade-off hypothesis is that this energetic balance has not become extinct in the modern working human. Rather, these mechanisms still operate but are triggered by different factors than in humans earlier. To explore this possibility, Laland, Levin, and Saturay project the working energy/take-home energy trade-off hypothesis onto the issues of job satisfaction and dissatisfaction developed in chapter 7. The result of this exploration is a series of predictions about the consequences of humans reacting to energetic trade-offs in postindustrial work and organizations. The most provocative of these may be the prediction that a "zone of conflict" exists inherent to the existence of a stable employer-employee relationship, in which increases in performance actually compete with increases in satisfaction, and vice versa.

In chapter 9, Steven Maier and Robert Levin develop the relationship between environmental challenges to animals, such as stressors and infection, and the work they perform. Maier and Levin begin by exploring responses to uncontrollable stress and by exploring research that shows the different reactions of animals to stressors in the presence and absence of control. Even small amounts of control can change the responses—a finding with implications for the human workplace. More directly, though, Maier and Levin show how responses to stressors in the presence and absence of control, mediated by different brain regions, affect the behavior of the animals in ways that affect both their energy conservation and their capabilities to seek out new sources of energy. Next, Maier and Levin explore the phenomenon of bidirectional control between immune system and brain in response to infection and explore the evolutionary developments that have led to animals utilizing the same system in responses to stressors. These responses also directly affect the ways that animals conserve energy, mobilize energy, and perform work. Maier and Levin conclude that these changes lead to important *intraindividual differences*—differences within the same individual between two different times—and that such differences, including in work performance, need increased emphasis, particularly in comparison with the emphasis on inter-individual differences on which such areas as employment selection and performance evaluation often depend.

The final portion of the book, Reflections, looks forward to where the integrative study of work in living systems can proceed from this initial exploration. As Reflections is the final portion in this guidebook, its emphasis is to note ways that the approaches developed for this volume can be extended to potentially fruitful disciplinary and interdisciplinary research. These possibilities are developed in several views: Perspectives on Exploring, Observations, Design Interlude, Work Integrates Life, and Life Integrates Work. The Reflections materials thus summarize perspectives gained from the exploration itself. The Reflections materials, including the Observations portion and the Design Interlude figure, can therefore also be used by readers to supplement the Introduction, when beginning the book or for perspective along the way.

The volume started by exploring how work can look different when explored from the perspective of life. We can see how this approach can help develop a richer perspective of work and its connections to life and to living systems. We can now also begin to explore life from the perspective of work—seeing ways that the various processes and perspectives developed in this book can help us better understand how the capability to do work is something central to life itself.

This chapter by Robert Levin brings together three major themes that can form a foundation for an integrative exploration of work in living systems. First, physical laws bound but do not determine how work gets done in living systems. Second, the chapter explores the fundamental processes by which cells perform work, whether in microbes or in humans, and shows how the characteristics of these processes, rather than those of physical limits, provide a foundation from which work at all other levels of organization proceeds. Third, the chapter describes the natural history of these work processes at a molecular level, taking us back to the earliest history of life on Earth and showing us how then work met life.

1 The Ancient Patterns of Work in Living Systems

Robert Levin

Work is a process or set of processes. Every process involves one or more steps, each one of which shapes and bounds what emerges from that process. Any biological (or nonbiological) work process requires energy, and the work that is done with energy produces change: change in the organism performing the work and change in whatever system on which the organism is performing work.

This chapter focuses first on one of the most fundamental—and surprising—ways in which energy and the capacities to induce change through work are connected in biological systems. The characteristics of this energetic process, like any process, fundamentally shape and bound what is possible and what is not possible in the work of any living system, forming patterns of work at different levels of organization in every organism—both the work that goes on inside each organism and the work that every organism does on the world, including the most sophisticated reaches of human work.

Many of the researchers (from fields within physics, chemistry, and biology) who originally helped to uncover this fundamental energetic process, to understand its centrality to the development and evolution of life, and to explain what they had found—first to other specialists in their own fields and then to the wider scientific community—were willing to set aside their own preconceptions, and others', too. For these researchers, this meant setting aside for instance, preconceptions about how chemical reactions power cells, how life developed and what forms life takes, and about what work is in a living cell.

To do so, they each challenged the veracity of substantial "placeholders" (that is, widely accepted but nevertheless unsubstantiated pieces of "knowledge") in their fields, placeholders that actively blocked progress toward a fuller, more accurate understanding of the way life turns energy into work. One of these controversies was routinely described at the time as a series of "wars": Researchers set aside their own and others' preconceptions and placeholders at great cost—personally and to their careers.

One of these researchers, Franklin Harold, devoted a great deal of his career to explaining the processes covered by this chapter, first in careful reviews that explained these discoveries to scientific specialists (Harold, 1974; 1977), and then in a book, *The Vital Force* (Harold, 1986), which explained these complex bioenergetic phenomena to biologists more generally. *The Vital Force*; an article he wrote providing more recent perspective, titled "Gleanings of a Chemiosmostic Eye," (Harold, 2001); and a series of conversations we were fortunate to have with Frank in 2004, through the good offices of Norm Pace; together form a strong foundation for this chapter, one from which we have drawn throughout, along with the work of others.

Our focus here is, first, to understand what thermodynamic and physical laws can tell us and cannot tell us about how work gets done in a living system. Second, it is to understand the work process by which all work fundamentally gets done in all cells and therefore in all living systems, a process known as *chemiosmosis*. Understanding this work process can provide a common grounding for our understanding in later chapters of how work gets done at very different levels of organization. Third, it is to understand how the origins of the enzyme that makes this work process happen is closely tied to the early history of life on Earth. In turn, this history allows us to consider more closely the role of work in early life and in the radiation of life, and to consider more closely the implications of continuity, rather than discontinuity, of basic forms of work across the broad array of organisms on Earth.

Of the possible ways to begin a book exploring work, tackling a potentially arcane cellular process may at first seem like an unlikely beginning. But this is a book that integrates the study of work across various levels, noting similarities, connections, and interactions among them. What better place to start than with the most basic, fundamental level at which work gets done in living systems?

Foundations for Work in Living Systems

First principles do not mandate that different forms of life need to work in the same way. Indeed, one of the themes of this book is that *work* is generally not one thing, but many things. Whether or not the study of work in living systems can be performed in an integrative manner has to do with the actual, empirical relatedness of these many things. First principles, such as thermodynamic principles, describe some crucial constraints about how living systems must work, but they do not mandate that these constraints would necessarily be met in similar ways in different organisms.

(Readers of the portion that follows are also referred to the descriptions of the first and second laws of thermodynamics, and other fundamentals for work in living systems, in the Observations portion in the Reflections section of this volume and to the graphical representation of energetic (and other) transformations and trade-offs in the Design Interlude portion of the Reflections section.)

Thermodynamics and Work

First Law of Thermodynamics   The first law of thermodynamics, known as the law of the conservation of energy, is essential for understanding how living systems work. Indeed, it is both the founding principle of the study of physiology and the contribution of physiology to thermodynamics (Atkins, 2007; Rabinbach, 1990). The law is stated in different forms in different contexts, but a form of the law that focuses on two aspects particularly related to work in biological systems is relevant to us here. That form states, first, that energy can be transformed from one form into another, which is a constant motif of how work gets done in living systems. Second, although energy can be changed into different forms, it is not gained or lost but always conserved. Thus, the energy that it takes for you to sit and read a newspaper is the same whether that energy is provided by crackers or carrots. The first law, conservation of energy, is responsible for the fact that the energy of crackers can be changed into the energy required to acquire a visual image, change the image into nerve impulses, and extract meaning from those impulses. The first law applies just as equally to nonliving systems, such as steam engines, as to you, the crackers, and reading the newspaper (Atkins, 2007).

In both living and nonliving systems, the basis by which this principle is put into practice is referred to as *energy transduction*. Biological systems have developed remarkable capacities for energy transduction that are embedded in biological structures (Harold, 1986): for example, to change crackers (to continue our example) into glucose molecules, into various forms of energy carriers, into concentration gradients that perform a remarkable amount of the work in the cell, into one form of those gradients, ion gradients, that allow the retina to signal the black and white edges of the type in the newspaper column and that also allow your brain to read the words, to understand the words, and to make the decision that the columnist is ignorant.

In addition, the law of the conversation of energy says that if you are reading your newspaper on the computer, the amount of energy required to power the computer is the same whether the power comes from you riding a bicycle-powered generator (e.g., Thompson, Foster, Eide, & Levine, 2008) or from a distant coal-fired power plant. No work by a biological system, even a system as complex as our own, can transcend this law (Brooks & Wiley, 1988).

At the same time, the first law does not compel biological unity. It does not require that you eat carrots, that a rabbit eat crackers, or that an *Aplysia* sea slug read the *Guardian*. Thus, the first law provides one fundamental boundary of what we will come in this chapter to call a *performance envelope* of how work gets done in a living system.

Within the first law, though, and any other such bounds we come to discover, any-thing goes. This envelope thus also provides a basis for the diversity of work we see in the natural world and in human work—diversity built on a particular biological foundation.

Second Law of Thermodynamics   Likewise with the second law of thermodynamics. Applied to biological systems, the second law states that, because the order in the universe is constantly decreasing, energy is constantly required in biological systems to maintain the order that is necessary to maintain life. This constant requirement for energetic inputs directly affects the work that an organism must perform on the world. It also affects the internal work that an organism must perform. DNA, for example, is a large, complex, highly ordered, highly structured, and highly improbable molecule. It exists "far from equilibrium" in the terms of the second law. Large amounts of energy are therefore required to synthesize even a single DNA molecule, and additional constant infusions of energy are required to maintain, repair, and utilize such a complex and highly ordered biological structure (Atkins, 2007; Voet & Voet, 2004).

An even plainer meaning of the second law is that if you sit in that chair reading your newspaper for long enough without eating any crackers, you will not be around any longer. You, like any other living system, need a constant infusion of energy to exist. The fact that you are built that way, by ages of evolution, has some surprising consequences for how you can and do work, even if crackers are in plentiful supply, consequences which are explored both in this chapter and in the treatment of various trade-offs that occur throughout this volume.

Third Law of Thermodynamics   Your ability to continue to sit and read your news-paper would become severely limited in a temperate climate, even with endless crack-ers, if the heating in your house were to fail in midwinter. This brings us to the third law of thermodynamics, which, for our purposes, says that temperature is a funda-mental characteristic of how work gets done (Atkins, 2007). This may seem irrelevant today, because when we sit working in front of our computer screens, we are organ-isms with a constant brain temperature and body temperature, in a room with a near-constant room temperature (save heating failures). But temperature has a direct relation with whether fundamental work processes in living systems can get done at all, a direct effect with perhaps the sharpest boundary on the performance envelopes in which we work (Boutilier, 2001; Raison, 1973). Later in the chapter, we describe what happens to the organism if that boundary is crossed and how that effect is tied directly to the question of how work gets done in the cell (Hochachka & Somero, 2001; Levin, 1984).

If these various fundamental performance boundaries seem particularly invisible to us today, they are by no means invisible to the overwhelming majority of organisms, whose body temperatures fluctuate with environmental temperature. The strategy of a constant body temperature, homeothermy, no doubt contributes to the smooth functioning of a complex brain. Yet it is energetically an expensive undertaking: The first and second laws of thermodynamics require that an organism maintaining a constant body temperature find energy sources to maintain constant body and brain temperatures in the face of the surrounding environment. So, too, with our homes and offices: If these are to be temperate when the environment is arctic, then the first law obliges that logs be hauled from a forest or coal hauled to a power plant to make it so. It is not a requirement of the first law that the work of the smokestacks and the turbines that, in turn, power the manufacture and the operation of our computers and that heat our offices be visible from our window so we can always be reminded that their work is inseparable from ours. However, their existence or the existence of some other power generator is a requirement (Atkins, 2007). Their work is not only inseparable from ours, but insuperable as well.

Each of the first three laws of thermodynamics thus provides us with important boundaries of the performance envelope within which work in living systems must be done. Yet nothing in the three laws requires or compels that there be some fundamental form of biological work. Each organism might perform its work in myriad ways. Indeed, when we look at all the ways in which organisms perform their work on the external world (Odling-Smee, Laland, & Feldman, 2003; Pace, 1997, 2001), that appears to be the case. We therefore want to explore further how organisms perform the work that goes on inside themselves.

Physical and Mechanical Work

Before delving inside the organism and the cell, let us take a closer look at the readily visible kind of work we see most often when observing an animal foraging, a plant growing upward against gravity, or a bacterium migrating up a gradient of nutrients, using its tail rotor to move and tumble in the general direction of more food. All of these are examples of physical work, and the description of work in physics or mechanics matches nicely our intuitive sense. (One reason for this is that the use of *work* in physics is a coinage that adopted the already existing common meanings of the word *work* in using it as a label for a particular technical sense (Rabinbach, 1990)).

In describing physical work, one can, for instance, imagine performing the work of rolling a rock up a gradient or hill. It is not surprising that if you roll the same rock up a higher hill, you do more work than if you roll it up a lower hill. Likewise, if you roll a small rock and a big rock up the same hill, you do more work rolling the big rock than rolling the small rock. You can also imagine, thanks to the law of the conservation of energy, that if you put the work into rolling a rock up to the top of a hill, the rock could also perform work rolling back down.

In fact, this kind of work, coupling two processes together, is done every day. For example, you would also be doing work if, instead of rolling a rock up a hill, you pumped water to the top of the hill. You can imagine that if you operated a hand pump at the bottom of the hill, you would do a lot of work to get it to the top of the hill. You could also use a pump that operated by electric current. In that case, the current would have to do exactly the same amount of work as you did by hand (Atkins, 2007).

Once the water was at the top of the hill, then just like the rocks, the water would have the *potential energy* to do work when coming back down again. But if you pumped the water uphill into a tank, and then just tipped the tank over, the water would be unlikely to do much *useful work* as it meandered down the hill in rivulets, with its energy dissipating among the rocks and grasses.

If, instead, you channeled that same tank of water down a water course, then it could do useful work. You could place a waterwheel at the bottom, which is simply one form of a motor. That current of water could turn the motor. The motor could do physical work, such as turning a generator to produce electrical current and thereby electrical power, to do work. In fact, this paradigmatic example of energy transduction (e.g., Lane, 2005) takes place every day in mountainous and coastal regions around the world: Work is performed at night to pump a source of water up a pipe—up a gradient—at a time when demand for electrical power is low. Then, during the day, water is released from storage to flow downhill and perform work on a waterwheel, driving a generator to produce electrical power. The law of the conservation of energy makes this possible. So does the principle of energy transduction that flows from it.

This same coupled mechanism for doing work is used to generate tremendous power within the cell: The understanding of how work gets done in living systems depended on the discovery that this is also how an organism does the most fundamental work that occurs inside the cell.

Discovering the Fundamental Work Process in Living Systems

The fundamental information process of life, based on DNA, has become widely known through the dramatic circumstances of the discovery of its structure in 1953 and its scientific and applied impact since that time. Yet the similarly fundamental energy process of life is less widely known and appreciated in comparison, even though the pursuit of this answer was just as sought after as a grail for many years, its discovery was not without drama, and the Nobel Prize was awarded for its discovery in 1978.

Despite the importance of DNA as a complex molecule that is a catalog of vital information for an organism and its descendents, information is only one half of what life needs. It also requires energy to live and to perform work (Lane, 2005). Moreover, without work and energy, there can be no functional biological information. Without energy, an organism cannot perform the work required to synthesize DNA in the first place, to transcribe DNA and RNA, or to synthesize proteins from the information in DNA and RNA. Moreover, DNA and RNA exist in an aqueous *cellular milieu* in which any number of proteins from any number of sources and causes affect what is read from the nucleic acids (Alberts et al., 2008). All of these processes require energy and work, without which the DNA, RNA, and proteins no longer can provide useful information. In addition, one of the most noticeable aspects of biological information processes is its diversity; one of the most noticeable aspects about the energetic side of life is the way that the energetic processes are conserved.

Inside the Cellular Performance Envelope

Early physiologists realized that energy was required for an animal to survive and to perform physical work: The law of the conservation of energy requires it. It was also obvious that for an animal, ingested food had to be the source of that energy. The question was just how foodstuffs came to power animal function. It was obvious that the answer would lie within the cell. As investigations by researchers in biochemistry and in the emergent field of bioenergetics proceeded, particularly in the last half of the nineteenth century and first half of the twentieth century, they discovered that energy was often obtained from organic nutrients, for instance glucose, produced from the breakdown of different foods. (Plants and a range of other organisms first obtain energy directly from the Sun via the work of photosynthesis, described in chapter 4, through which they convert this energy into organic nutrients.) It was also clear that it was not possible for the cell to "burn" glucose directly to power muscle contractions or other energy-requiring processes (Harold, 1986; Lane, 2005) (Here, and with regard to the discovery of ATP below, and its synthesis as well, we commend our readers to excellent and more detailed historical accounts of early and more recent bioenergetics research that can be found in Harold, 1986, and Lane, 2005).

There are four requirements in order for useful work to be done in a cell using glucose (or other organic nutrients). The cell must have:

1. A way of breaking down the glucose or other energy source step by step, rather than burning it all at once, so that some of the energy can be captured as work and not wasted as heat;
2. An energetic intermediate, a carrier molecule, that can effectively carry (i.e., transduce) energy between the energy source (e.g., glucose) and the processes that require energy (e.g., making muscle or DNA);
3. A way of making or synthesizing carrier molecules; and
4. A way of generating the energy to make the carrier molecule—an energy expenditure required by both the first and second laws of thermodynamics.

Cells use many so-called energetic intermediates or carrier molecules. These make possible the energy transduction required for energy from food sources to be able to be applied to the countless essential cellular processes that require energy. There are about a dozen well-characterized energetic intermediates in the cell. The primary—but by no means sole—intermediate is the ubiquitous, relatively simple compound known as ATP—adenosine triphosphate.

For many years, the scientists who discovered ATP and did much of the early research on its function, such as Nobel Prize recipient Fritz Lipmann, believed strongly that ATP was able to perform its energy carrier function due to some special properties of its bonds, properties for which they and others then searched tirelessly but in vain. Around the same time as the initial discovery of ATP by Lipmann, in 1941, a number of researchers working on metabolism, including Hans Krebs, who discovered the citric acid cycle in 1938, and Hermann Kalckar, who discovered oxidative phosphorylation in 1940, found various processes within a cell that released energy from foodstuffs in a stepwise way that coupled each step with performing useful work (Harold, 1986).

Each of these steps involved a chemical process known as oxidation. The use of the term *oxidation* stems from early chemistry experiments into the process of obtaining energy from foodstuffs and similar substances that did indeed involve burning. With modern biochemistry, we now know that oxidation transfers energy from food sources by pulling electrons off of food source molecules, such as glucose, for electrons have high amounts of energy. "Oxidation," thus, in our context, refers to the process of electrons being removed from food molecules (or other molecules) sequentially, in order to power other, energy-requiring processes (Hill & Kolb, 2007; Lane, 2005).

The researcher who fully worked out, during the 1940s and 1950s and in the face of much controversy, the individual steps by which oxidation yields a step-by-step flow of electrons that can be used to power other chemical processes, was a biochemist named David Keilin (Harold, 2001; Mitchell, 1979; Lane, 2005). Keilin specifically showed how, in each step, enzymes removed electrons from foodstuffs, and thereby created a chain in which the energy from the electrons could be used to drive steps in metabolism that were otherwise energetically unfavorable and would not occur on their own (Alberts et al., 2008).

Each part of the process eventually yielded to scientific progress and technical advancement, except for the final crucial step. In this step, the cell actually had to manufacture or synthesize the key energetic intermediate, ATP, the energy carrier. Because manufacturing ATP requires energy, both the chemical reaction and the energy source for this work had to be found as well. Efforts to find the missing chemical reaction for synthesizing ATP or its source of energy for a long time came to naught. Then, in the early 1960s, Peter Mitchell (who is both the Watson and the Crick of the adventure that follows) set off an uproar by proposing that there was no chemical reaction at all.

Physical Work Powers Cells: Discovering Chemiosmosis

"The obscure we see eventually, the completely apparent takes longer."
—Peter Mitchell

Mitchell overturned the card table by proposing that neither ATP synthesis nor its source of energy came from a chemical reaction. Instead, he claimed that the cell did this essential work physically, not chemically. He also claimed that the power to perform this physical work came from David Keilin's controversial electron transport chain (Lane 2005; Mitchell, 1979). Mitchell proposed that these essential processes were performed by two coupled forms of physical and mechanical work. Both involved the work related to moving hydrogen ions, or protons, across an impermeable membrane: One form of work, called *proton pumping*, moved the protons in one direction (e.g., from the inside to the outside of a bacterial cell membrane), and the second, coupled form of work, *ATP synthesis*, harnessed the energy from the resultant flow of protons back into the cell. Mitchell saw the work done to generate and then utilize the flow of protons to be the essential work process in the cell, coupling proton pumping to ATP synthesis, and he coined the term *chemiosmosis* to describe the flow of protons and the work that was done (Harold, 2001; Mitchell, 1979).

In proton pumping, the energy from the electrons that have been extracted from the stepwise breakdown of the sugar is used to pump protons—hydrogen ions, H^+, the ion of the simplest and most plentiful molecule in the universe—across the lipid bilayer membrane that surrounds the cell of a bacteria (or an archaea, see below), and which forms the mitochondrial membrane in eucaryotes like ourselves. Because a lipid bilayer membrane is largely impermeable to the passage of a charged ion, such as a proton, it requires actual work to pump, or *translocate*, the protons across the membrane. This work is fueled by the energy obtained from the electrons removed from foodstuffs during stepwise oxidation. Just as it would take work to stand at the bottom of the hill and pump water up a gradient, it takes work to pump protons across a membrane and up a gradient: When you pump protons, you create a concentration gradient with more protons on one side of the membrane than the other. The more protons you pump, the more work you have to do because you are increasingly pumping against the concentration gradient (Alberts et al., 2008; Harold, 1986; Harold, 2001; Lane, 2005; Mitchell, 1979). As you do so, you are also storing more potential energy on the other side of the membrane.

One fundamental kind of work in the cell, therefore, is proton pumping. But, just as with pumping water up a gradient, pumping protons up a gradient and creating potential energy does not necessarily result in any useful work in return. If water leaks back down a hillside or if protons were able to leak back through the membrane, none of the potential energy built up would be turned into useful work. However, if a channel directed a current of water—or a current of protons—through a turbine, it could create a powerful electrical charge.

This is what happens to the protons in chemiosmosis. The lipid bilayer membrane remains impermeable to protons (and to other charged ions used in some chemiosmotic work, such as sodium (Na^+)). The protons pumped across the membrane cannot leak back across. The concentration of the protons on one side of the membrane remains higher, with two effects. First, this concentration gradient is capable of doing physical work, proportionate to the magnitude of the gradient. Second, protons, H^+, carry a positive electrical charge. The lipid membrane has low conductance and is well suited to the buildup of a charge in the presence of a concentration gradient of charged ions. Thus, an electrical charge also builds between the increasingly positive side to which protons are pumped and the increasingly negative side from which protons are pumped (Harold, 1986).

The combined physical and electrical forces generated by the proton buildup were referred to by Mitchell as "the proton-motive force." The potential power of the charge that develops is extremely large, in part because, once we find a way for the charge to get across the lipid bilayer, it is going to be discharged across only the miniscule thickness of the membrane and a cross-sectional area of a channel through the membrane about the size of a proton. Over such a minute distance and area, the power is proportionately as great as that contained in a bolt of lightning (Lane, 2005; Nicholls & Ferguson, 1992).

ATP Synthesis from a Proton-Charged Turbine

Peter Mitchell thought that a charge of that magnitude could certainly do large amounts of useful work if directed in the right way, and that the right way would be the second of the two forms of coupled work: ATP synthesis. Mitchell hypothesized that this charge was of sufficient magnitude to bring about the synthesis of adenosine triphosphate (ATP) from adenosine diphosphate (ADP) and inorganic phosphate (P_i) (Harold, 2001). Mitchell also hypothesized exactly how the charge would be directed to do the useful work of synthesizing ATP: When the protons are initially translocated, charge builds up on one side of the membrane. On the other side sit quantities of ADP and P_i, waiting to be synthesized into ATP. Mitchell reasoned that, because the membrane itself was impermeable to protons, there must exist a portal, a way for protons to flow back into the cell. The portal had to provide a way to complete the circuit, to harness the immense energy available in the charge, and to use the energy specifically to make ATP from ADP and P_i. He therefore concluded that the portal was an enzyme, a protein catalyst, embedded in the membrane and with a structure that would (a) serve as a channel or gateway for the protons to flow back into the cell and complete the circuit, (b) capture the energy of this flow of protons, and (c) use it to manufacture ATP (Harold, 1986; Mitchell, 1979).

It is easy to miss the comprehensive breadth of Peter Mitchell's hypothesis. Mitchell hypothesized, all at once, that:

1. The power for synthesizing ATP would come by pumping protons across an impermeable membrane, rather than from a chemical reaction;
2. The protons would create an electrical charge;
3. The force of the proton buildup, the proton-motive force, would have to be channeled through an enzymatic turbine that had to be embedded in the bilayer membrane; and
4. The enzyme that served as a channel would also synthesize ATP and would do so by using the power of the proton circuit.

And he was right. Indeed, when the actual structure of Mitchell's hypothesized enzyme, called *ATP synthase* by standard enzymatic nomenclature, was discovered some years later, it turned out to perform the functions Mitchell hypothesized, and Paul Boyer, John Walker, and Jens Skou shared the Nobel Prize in 1997 for discovering the structure of the ATP synthase enzyme (Harold, 2001; Lane, 2005). ATP synthase is built in two parts: One part (called the *rotor*) spans the membrane and provides the portal hypothesized by Mitchell for the protons to move back into the cell. This part also harnesses the potential energy of the proton lightning bolt to do useful work: The energy of the protons passing back into the cell induces mechanical changes, referred to as conformational changes, in the first part of the enzyme (Walker, 2009; includes an animation of the enzyme rotating at some 150 revolutions per *second* within the membrane). These mechanical changes and the available energy from them is transmitted by the rotorlike structure of the first part of the enzyme, which, in turn, induces changes in the second part of the enzyme (called the *stator*), which is tightly physically bound to the first part (Alberts et al., 2008; Boyer, 1997; Leslie & Walker, 2000; Walker 2009).

The second part of the enzyme, which is bound to the first part of the enzyme but is not membrane-bound, therefore undergoes changes driven by the rotor. The second part successively:

1. Opens binding sites for an ADP molecule and for an inorganic phosphate molecule (P_i);
2. Brings those sites with their molecules physically together, so that ATP is formed from ADP and P_i; and
3. Releases the synthesized ATP molecule (Boyer, 1997; Stock, Gibbons, Arechaga, Leslie, & Walker, 2000; Stock, Leslie, & Walker, 1999).

The rotor is thus a motor in the literal, not the figurative, sense. The motor requires a set ratio of protons that must pass through and drive the ATP synthase motor to create each molecule of ATP synthesized. Generally, three protons are required for the synthesis of one molecule of ATP (Nicholls & Ferguson, 1992, 2002). ATP is thus manufactured by a motor driven by a turbine, and turned by an electrical current.

In chemiosmosis, then, two forms of useful physical and mechanical work, proton pumping and ATP synthesis, are tightly coupled together (figure 1.1). The ATP that is produced allows other forms of work to take place in an organism, such as protein synthesis, muscular work, and the transmission of information from a sensory organ to a brain. Work in living systems therefore has a specific and fundamental biological foundation. For anything to get done in the organism—to move a muscle, to sense one's surroundings, to utilize DNA—each of these requires a certain number of turns of the crank of the ATP synthase, which, in turn, depends on a certain number of protons passing through the enzyme's channel. The rate at which any other cellular or biological work can get done depends upon this rate of ATP production from the work of the proton circuit. Because all other work depends on, and draws on, this process, energetic demands are constantly trading off against each other, as different kinds of work are done in the cell or in the organism, the body.

The Work of Concentration Gradients

Other forms of work in the cell follow a similar motif as chemiosmosis, in particular by building up concentration gradients that can perform work. The mechanism of having a simple system perform the work of creating a concentration gradient is a repeated motif by which useful work gets done in the cell. Building concentration gradients is therefore a basic form of work in biological systems. For instance, this basic form of work is how ATP performs its function as an energetic intermediate. In fact, the realization that ATP functioned through concentration gradients, rather than by having special properties of its bonds, was yet another consequence—and yet another controversial one—of Mitchell's hypothesis and research. As two more examples, the work of pumping charged ions to create concentration gradients is a basic approach to storing and transmitting information in biological systems (chapter 2); chloroplasts use proton gradients and chemiosmotic processes as they harvest energy from light (chapter 4).

Figure 1.1 (facing page)
The big picture of the work of chemiosmosis. Processes described individually in the text are shown together in this schematic: Proton pumping across the membrane; the flow of protons through the ATP synthase portal; and the proton-driven changes driving the schematic ATP synthase rotor against the stator, catalyzing the synthesis of ATP from ADP and P_i, and resulting in the buildup of concentrations of ATP.
Original illustration by Alan Blackwell. (Selected elements after Harold (1986). Readers desiring illustrations of individual processes depicted here are referred to such sources as Alberts et al. (2008).)

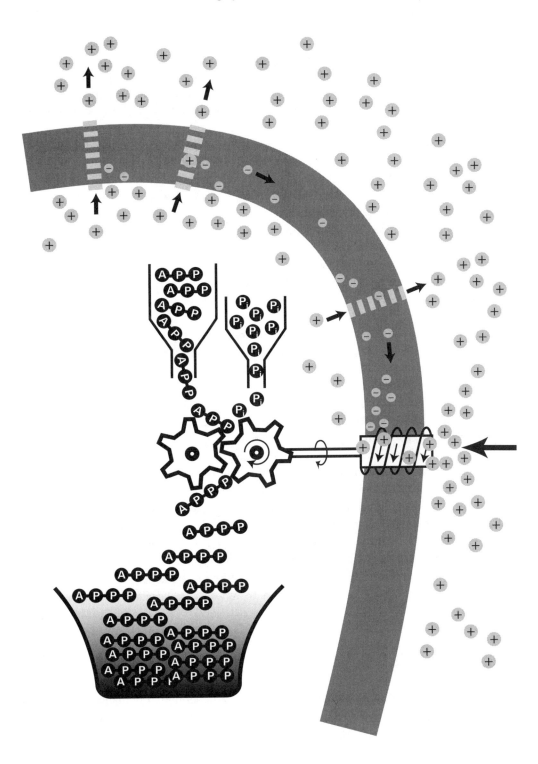

Concentration gradients in turn follow basic chemical and physical laws determining how much work it takes to form and maintain them and how much energy can be extracted from them. In general form, the physical work that can be done by a concentration gradient increases with the magnitude of the gradient. Likewise, for charged ions, the magnitude of the electrical charge is, in many cases, also proportional to the gradient. (These effects are described more precisely in the Nernst equation (e.g., Harold, 1986).) Therefore, the capacity of an organism to do some complex work—for instance, to receive information, process it, and do something useful with it, one fundamental pattern of work—is not merely a matter of interest, motivation, or will but is shaped and bounded by the physical capability of the system to do the work necessary to develop and maintain the myriad concentration gradients that keep the organism alive, sensate, and working—inside and out.

ATP Synthesis, Concentration Gradients, and the Consequences of Work in the Cell

ATP performs its own work as an energetic intermediate by a similar process for concentration gradients more generally. A high concentration of one molecule relative to another creates a potential that can be used for work, as in the proton gradient created by proton pumping. The amount of potential work is proportional to the ratio between the two molecules, a fact crucial to the way that ATP performs its role as an energetic intermediate (Alberts et al., 2008; Harold, 1986). The cell regulates the balance of the reaction between ATP and ADP + P_i from moment to moment, always in the direction of building and maintaining a high concentration of ATP (Atkinson, 1965; Hardie & Hawley, 2001). In the functioning cell, there is a high concentration of ATP relative to the concentration of ADP, roughly 10:1 (Beis & Newsholme, 1975).

With this high ratio of ATP to ADP, ATP naturally runs "downhill" to ADP + P_i readily, and this energetically favorable reaction can be coupled to drive innumerable cellular processes that are energetically unfavorable—and essential to life. If more energy is required in a given process to be driven, ATP can actually run further downhill, directly from ATP all the way to AMP (adenosine monophosphate) (Atkinson, 1965; Harold, 1986), because the ratio of ATP to AMP is usually on the order of 100:1, and, in some muscles of some species, it is as high as 400:1 (Beis & Newsholme, 1975). This substantial gradient allows a great deal of energy to be generated quickly when need be. It also requires a great deal of energy to synthesize ATP again after it has been "degraded" into AMP.

Because ATP will spontaneously run downhill to ADP and AMP, it takes a lot of energy and work to maintain the energy charge by keeping the cell stockpiled almost entirely in ATP. As long as the high concentration of ATP relative to ADP is maintained, this gradient is always available to perform work in the cell. However, the amount of work that ATP can perform is entirely dependent on maintaining this concentration gradient. If ATP and ADP are instead present in about equal proportions, there is almost no energy available to do work (Alberts et al., 2008; Lane, 2005). In fact, cells—and therefore unicellular organisms—regulate their activity levels according to the ratios of ATP, ADP, and AMP, so that activity is instantaneously reduced when relative ATP levels fall (Chapman, Fall, & Atkinson, 1971; Hardie & Hawley, 2001).

Thus, cells are working constantly to maintain a high relative concentration of ATP. Cells are thus always working relentlessly, in the literal meaning of that term. When you, as an individual human, take time to think, a tremendous amount of energy is required for that time. While you are thinking, your cells are still constantly grinding out work via chemiosmosis to generate quantities of ATP. The magnitude of work that must be invested in ATP production for any organism to survive and to perform other work under such conditions is astounding. For instance, ATP synthesis has been measured in bacterial species at higher rates of energy production per unit mass than occurs in the Sun (Harold, 1986). This startling fact is made possible in part because the Sun is not very dense, and a bacterium is as dense as water, but it is still a remarkable rate of energy production. As another measure of the magnitude of this work, ATP turnover in a human is on the order of a human's body weight: (about 65 kilograms (140 pounds)) of ATP each day (Nicholls & Ferguson, 1992; Lane, 2005; Walker, 2009).

This research explaining how work is done in the cell, how much work has to be done on a constant basis, and the central role of the proton circuit and of the membrane-embedded ATP synthase all come together to explain one consequence at which we hinted in discussing the importance of temperature to performing work. These all combine to explain both how and why organisms die of the cold. One might think that this happens simply because chemical reactions run slower at colder temperatures. But at the mildly cool temperatures at which humans, for instance, die of cold, that difference is not significant enough to create such drastic consequences.

Instead, an organism or cell without enough energy starts to cool to the temperature of its environment. When it passes a fairly mild critical temperature (around 25°C (77°F) in humans), the bilayer membrane undergoes a physical effect, a phase change. In water, this might be from liquid to solid. In bilayer membrane, much like in molten candle wax, this is roughly from liquid sludge to more solidified sludge. That phase change kills the chemiosmotic circuit in both directions: The phase change produces "cracks" in the membrane, short-circuiting the proton flow, and the solidification of the membrane deforms the ATP synthase enzyme, embedded in the membrane, halting ATP production (Thurston, Burlington, & Meininger, 1978). The cell dies, and the organism follows soon enough (Levin, 1984; Raison, 1973) This effect is powerful enough to create selective pressures in fish, for example, so that small changes in DNA sequences convey large changes in the temperature range to which their enzymes are adapted (Hochachka & Somero, 2001).

One other form of physical work, present in bacteria, relies not on ATP, but solely and directly on proton pumping: a naturally occurring demonstration that the ion-pumping processes we have been describing really perform work in the everyday sense of the word. Bacteria, such as *Escherichia coli*, locomote through a nutrient gradient, in a laboratory or inside your gut, using their flagella as rotors or propellers. The flagella rotate first in one direction and then the other, generating two swimming states, in which the cell either (a) moves generally in a straight line or (b) tumbles randomly. A three-dimensional random walk thereby results—at least when nutrients in the environment are evenly distributed. But in a concentration gradient of nutrients (such as might exist, for instance, in a pond or in the human gut), the chemosensory pathway of the bacterium shortens the tumbling steps *down* the gradient, with the net result that the bacterium executes a biased random walk *up* the gradient toward the nutrient source. The flagellar motors propelling these movements are, however, not powered by ATP, but directly by proton pumps, which therefore not only do work on the inside world of these bacteria, but on the outside world as well (Bray, 2002; Falke, Bass, Butler, Chervitz, & Danielson, 1997; Manson, Tedesco, Berg, Harold, & Vanderdrift, 1977). (Using hydrogen fuels to power transportation is evidently not a recent innovation.)

Having explored the world of work of chemiosmosis, it is now time to explore its natural history and what that history tells us about work and life.

The Molecular History of Work

Our exploration of chemiosmosis gives us an understanding of a basic biological work process. We can see how this work process can form a widespread foundation for work in living systems, with more complex forms of work—the myriad processes in the cell as well as the different kinds of foraging used by species—built on top of this basic mechanism. Once chemiosmosis was identified as a fundamental mechanism for energy and work within some experimental organisms, a host of questions came to occupy researchers from a range of fields: How widespread is chemiosmosis? Does it occur in all forms of life as we know it? Is chemiosmosis a relatively recent innovation? Or did it appear early in the development of life—and work—on Earth? The answers to these questions came in a somewhat unexpected form, from different research groups, and turned out to provide novel findings central to the understanding of the early history of life—and, for our perspective, the early history of work.

Researchers answered these questions by unraveling the molecular heritage of these work processes, by developing a *molecular phylogeny* of the molecules that make up the chemiosmotic machinery. A phylogeny is a map of the relatedness of organisms over time. Traditionally, phylogenies were constructed as "trees" in a hierarchical or "progressive" relationship, also fitting with a view of species development that placed humans at the top (figure 1.2).

A molecular phylogeny compares similarities and differences across species between an individual molecule, such as a particular protein or genes for a specific type of RNA, and then uses these to infer similarities and differences between organisms and their relatedness (e.g., Alberts, 2008). These results are also represented as a *phylogenetic tree* that shows these relationships. The nature of molecular phylogenies is that they lend themselves to being displayed in a phylogenetic tree that "radiates" (figure 1.3). The result is a view of life that is consistent with the radiation of life on Earth and that is displayed without hierarchies.

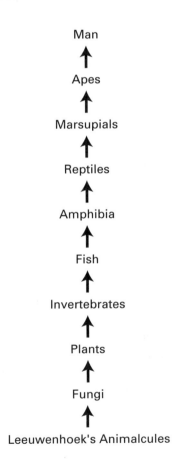

Figure 1.2
Example of a traditional hierarchical tree of life.
Courtesy of Norman Pace. Illustration credit: Alan Blackwell.

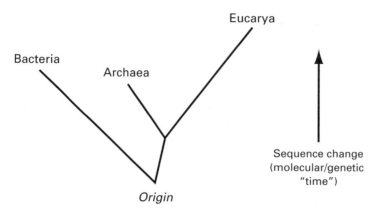

Figure 1.3
A simple radiating phylogenetic tree, in this instance representing the three domains of life described in the text. The arrow shows that increased radiation of the tree represents increased change in molecular sequences, described as molecular (not chronological) "time."
Courtesy of Norman Pace. Illustration credit: Alan Blackwell.

Although molecular phylogenies do not measure time (instead, they measure changes in molecules), phylogenetic trees can be used, often comparing among different molecules, to construct evolutionary histories of organisms, of molecules, and of processes with which the molecules are connected. Thus, ATP synthase, as an enzyme, is a protein, and molecular phylogenies can be constructed of ATP synthase—and have been. Because chemiosmosis requires ATP synthase, the molecular phylogeny of ATP synthase also provides us with a molecular natural history of chemiosmosis—across species and across evolutionary time. Delving into the molecular heritage of work can thus help us understand the work process, the diversity and unity of work among species, and the role this work process played in the emergence and early evolution of life.

It is useful to start by recognizing the broad range of diversity of species and the scale of time we are dealing with. In everyday life, we come across a small number of plant and animal species and think of those as representative of the species around us doing work. But, in fact, the vast majority of species on Earth are microbial. The plants and animals that are comfortably close to us occupy only small segments of the trees of our living heritage, based on the techniques of molecular phylogeny (Pace, 1997). Whether close or distant on the tree of life, we and they all do the same work of taking electrons from an energy source and by the same fundamental processes.

The other necessary perspective is that the span of time that living systems have been working since the emergence of life and the amount traceable through molecular phylogenies is unfathomably longer than scales we are used to dealing with in human history or indeed in human evolution. Life arose and had begun working in a way recognizable to our cells today by 3.5 billion to 3.9 billion years ago (Nisbet & Sleep, 2001; Pace, 2001), a factor of more than 2,000-fold greater than the time span since tool use began among human ancestors and more than 300,000-fold greater than the span of recorded human history. And it is a factor of some 350 million-fold greater than the length of time that something like, say, the cell phone or the laptop has been considered as an "indispensable tool" for performing modern human work. If one purpose of studying work in living systems is to provide a fresh perspective on work, then exploring the molecular natural history of work may help us to step back from our laptops for a billion, or three billion, years or more and get some of that fresh perspective. What is most unexpected is not how much looks different with this perspective but how much looks the same.

The Natural History of ATP Synthase and the Origins of Life and Work

To answer basic questions about the origins of chemiosmosis, molecular phylogenies were developed to compare the form of ATP synthase in eucaryotes—specifically, in the heart muscle of cows—with those in bacteria (Gogarten et al., 1989; Gogarten, Starke, Kibak, Fishman, & Taiz, 1992). The purpose was to see how closely related or conserved the ATP synthase would be in these two very different kinds of organisms. During the same period that this research was being conducted, though, a third previously unknown domain of life was discovered by other researchers using a different molecular phylogenetic technique: comparing differences in sequences of genes for ribosomal RNAs (rRNA) (Pace, 1997; Woese, 1998). This third domain of life was labeled Archaea (Woese & Fox, 1977). All of the archaea are microbial (some had been previously labeled as bacteria); many, though by no means all, archaeal species inhabit niches that as humans we label extreme environments, such as those at very high temperatures or pressures (Robertson, Harris, Spear, & Pace, 2005) (though there is no evidence that the archaea themselves find such environments extreme). One of the questions that arose immediately from the discovery of the archaea was how these organisms were related to the bacteria and to the eucaryotes.

As a result, the researchers working on molecular phylogenies of ATP synthases added samples of ATP synthases from archaeal species to their research. Their results, when combined with the results of the researchers working with rRNA, were a surprise, bringing at once new perspective on the origins and early history of life and on the origins of work in living systems, showing the truly ancient nature of ATP synthase and the chemiosmosis process, and showing the unity of this process across all forms of life and across billions of years.

The surprising result was this: ATP synthases obtained from archaea were more closely related to those from eucaryotes—to those from the heart muscle of cows—

than to those from bacteria. Over 50% of the protein sequences matched, or were *conserved*, between archaea and eucaryotes (even though the chemiosmotic processes in certain species use other charged ions in place of protons). That was startling enough. Even more startling, the sequence similarities indicated that ATP synthase had to be present prior to the divergence of Eucarya and Archaea (Gogarten et al., 1989; Gogarten et al., 1992; Woese, 1998).

There were many implications of this finding that became central for the history of life on Earth—and for the history of work in living systems. First, the conclusions of this interwoven set of research from different research groups indicated the presence of a last universal common ancestral state that existed prior to the divergence of the three domains (figure 1.4). Second, it showed that ATP synthase, and thus the work of chemiosmosis, had been present in this ancestral state—and for some time prior to the divergence. By combining phylogenetic research with evidence from early fossils, the divergence of the three domains from the common ancestral state is often placed as occurring 3.5 billion years ago, or somewhat earlier (Nisbet & Sleep, 2001; Schopf, 1999). Third, the existence of ATP synthase before the divergence from the last common ancestral state shows the ancient origins of this fundamental method for performing work—and its persistence over more than 3 billion years. Chemiosmotic work was thus present as far back in evolutionary time as we are capable of looking. (We should also note that one of the implications of the initial research on the early origins of ATP synthase is that its existence prior to the divergence of the last common ancestral state leaves open the possibility that ATP synthase was present prior to membrane-bound (i.e., cellular) life. The implication is bolstered by more recent research indicating that a number of bioenergetic enzymes existed prior to the emergence of cellular life, apparently performing their work in a precellular biotic environment (Baymann et al., 2002).) All of these findings, together, can truly change one's perspective of the centrality of biological work to understanding the emergence of life and to the ways that work gets done in all arenas today.

These same results show the universal nature of chemiosmotic work across all forms of life and simultaneously the diversity and unity of biological work. Eucaryotes, bacteria, and archaea all perform their basic work through chemiosmosis. The patterns of work thus created cross all divisions on the tree of life.

These patterns of work in the cell become the more remarkable to our understanding of work when paired with the broad diversity of patterns of work on the world performed by organisms across the tree of life. Each species performs its work on the environment, on its niche, in a somewhat different way. In the same way that niches of some archaeal species dwelling in the pores of super-heated rocks of volcanic vents have been defined as thermodynamic niches (Spear, Walker, McCollom, & Pace, 2005), we can ask whether niches could more generally be described in part by the work that a species performs to inhabit it. (Such an approach would

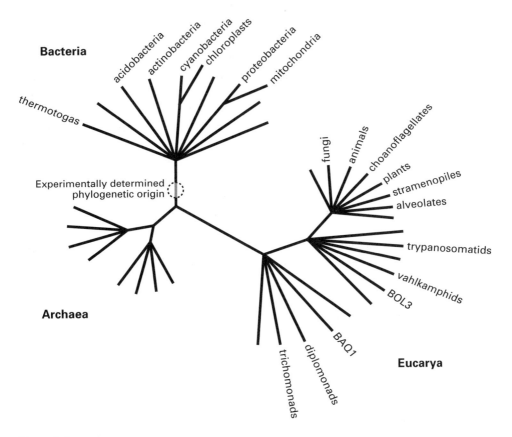

Figure 1.4
A radiating phylogenetic tree depicting the radiation of the domains Archaea, Bacteria, and Eucarya, and therefore at times referred to as the *Big Tree of Life*. The tips of the trees represent experimentally known starting points. The labeled *origin* region is developed from the experimental results of the molecular phylogenies developed from the starting points.
Courtesy of Norman Pace. Illustration credit: Alan Blackwell.

also be consistent with energetic definitions of evolutionary fitness described in chapters 5 and 8.) This diversity of work is built on a foundation of chemiosmotic work. Very little biological work, and no complex human work, can be done that does not reflect the patterns and rhythms brought about by the process of chemiosmotic work.

In this way, the work processes we have explored play two important roles in an integrative approach to exploring work in living systems. First, it grounds the thermodynamic and physical principles we described at the outset of the chapter within a fundamental and universal biological process. Work performed by a biological system, including the informational work described in chapter 2, passes through this gateway. This means that we can explore work in living systems without needing to resort in each instance to thermodynamic or physical first principles, but instead we can start from an empirical foundation in this process. Second, this work process can serve as a touchstone for understanding how work gets done. We said earlier in the chapter that no biological work can get done that transcends the law of the conservation of energy. Nor can work get done in a living system outside of the bounds of the performance envelope created by the nature and characteristics of chemiosmotic work. It is in that way a touchstone to which we can return in understanding work.

Implications for Exploring Work and Life

We have explored a fundamental process for performing biological work in living systems. The existence of the work process of chemiosmosis shapes our understanding of how work gets done in living systems. The laws of thermodynamics put bounds on how work can be done in the world. Thermodynamic laws do not, however, dictate that there need be a single, universal way that living systems perform work, though such laws do dictate what such a process can do and cannot do. The presence of an ancient and fundamental process such as chemiosmosis is instead purely an empirical and a biological reality.

Chemiosmosis is a process that can therefore provide one foundation for our exploration of an integrative study of work in living systems. With a process like chemiosmosis as fundamental, we can speak more surely of an exploration that is biologically integrative. The work of transmitting information and the work of photosynthesis, described in chapters 2 and 4, are phenomena that integrate directly with the work of chemiosmosis. When other contributors to this volume ask how organisms come to shape their niches or how designers can conceive and execute a sophisticated design, we will be exploring different boundaries of complex work. These boundaries, and the variability within them, are integrated as well with the boundaries of the performance envelope defined by the capabilities of chemiosmosis, by the physical forces of protons pumping across a membrane, and by the rhythms and patterns of work those forces create. One of the rhythms, for instance, is between the production of energy and its storage. At a basic level, this rhythm is contained in the flow of chemiosmosis and in the balance between ATP, ADP, and AMP, closely regulated by the cell.

The ways of working we are exploring have unfolded across evolutionary time. One view of the early microbial "workplace," prior to the divergence from the common ancestral state, is that early molecular evolution took place under increasing resource scarcity (or we might say within performance envelopes whose constraints were tightening), so that small changes in the capacity to do work more efficiently, and thereby utilize smaller amounts of nutrients and less ATP, had very high selective pressures (Alberts et al., 2008).

It is also possible to envisage a closely packed population of simple unicellular organisms, perhaps inhabiting an ultrahot deep sea vent or a volcanic heat vent, with most of the work the organism requires done for it by the surrounding heat (Pace, 2001). How early did these arise? We do not know. And how long before the growth in a population might have crowded individuals out into slightly colder environments? We do not know. But just a small amount of change would have required an increase in the amount of work that needed to be performed. We do not know how long life may have paused there before developing some new variation of a way to do work, making it possible to inhabit a new environment, a new niche. But we know that, at some point, a new way of working on top of chemiosmosis would have provided an organism with access to a new place in the physical environment.

The radiation of life, and of work, onward from there can be viewed in many ways. We choose to view it as a radiation of new means and mechanisms for performing work in the environment, building on top of chemiosmosis, on top of the work of pumping protons that was now locked in by evolution. Each new species, each new niche, then, represents a new way of working in the world, and we can come to see niches in that light. Among microorganisms living in a hydrogen-based environment within the microniches of geothermally heated rocks, it is possible to define organisms' niches thermodynamically (Spear et al., 2005). We can ask whether describing niches in terms of the work that an organism does to inhabit it could therefore be more broadly useful. In this way, the radiation of species can in part be seen as a radiation of different ways of working, allowing organisms to inhabit new niches. In this view, the tree in figure 1.4 is thus also a tree of work, of different forms of biological work on Earth. The radiation of work parallels—and makes possible—the radiation of life.

Viewing chemiosmosis as a fundamental way of performing work, common to all living systems, may also cause us to reexamine our understanding of human work and its relation to work in other species. It is understandable to view human work in the same manner as the traditional "progressive" phylogenetic trees and place human work at the most complex summit of a hierarchy in which the animals that perform work most like us come just below the summit, followed by organisms performing apparently less and less complex work. Our exploration in this chapter, and throughout this book, proposes another view: that of a continuing radiation of different forms of work among species and niches, paralleling the radiation of life, and making the inhabitation of new niches possible.

As one consequence, our perspective emphasizes the continuity between work in human and nonhuman living systems rather than the discontinuities. We believe that, throughout this volume, this perspective opens up new country to explore: Research on blowflies and on photosynthetic algae, for example, may give us insights into work that are more difficult to obtain from observing postal workers or primates—and vice versa. Utilizing knowledge gained from systems throughout the tree of life—or the tree of work—can help us to better understand a wider range of aspects of work in living systems.

Exploring these continuities can also help us move beyond distinctions drawn between "biological" and "psychological" approaches. Such distinctions fade into abstraction in chapter 2, for instance, in which the apparently psychological and apparently biological are combined, as such topics as energy, information, the work of the brain, and the effect on behavior are all investigated together. And not—at the beginning of the chapter at least—in the work of the brain of the human but instead in the work of the blowfly. If chapter 2 does not encourage these distinctions to fade, then surely chapter 3, which is based on research on the work of a nonliving system— manufacturing of computer chips—might help these distinctions to dissolve more completely.

From the perspective gained from investigating a fundamental work process in living systems and from coming to understand its origins, we can therefore see the apparent complexity of human work as one more flowering of the diversity of working processes among living systems. The diversity is built upon a simple foundation: energy used to do the work of moving the simplest ion in the universe, moving it a short distance, again and again, until the power of a lightning bolt develops, and work meets life.

Acknowledgments

Simon Laughlin, Christina De La Rocha, and Alan Blackwell each contributed content to this chapter well in excess of their respective editorial roles. The concept for the radiation of work with which this chapter concludes was developed from three separate sets of discussions with Christina De La Rocha, Alan Blackwell, and Norman Pace, respectively.

The illustrations for this chapter were developed with much time, thought, and creativity by Alan Blackwell. In addition, Alan helped work out how to meet the challenge of the development of the explanation of chemiosmosis and then helped with doing so on a practical basis, an effort on his part that made completing this chapter possible. Joseph Falke contributed both perspective and content that was always both timely and helpful.

Franklin Harold took the time for several extended conversations on the work of chemiosmosis and to explain the history of the research on chemiosmosis, in which he participated centrally, and which he has described in his scholarly papers and books. As an experienced scientist and thoughtful user of words in his writings, he was also perhaps our sharpest interlocutor on the meaning or meanings of the word *work*.

We acknowledge Nick Lane and his treatment of chemiosmosis and related topics in the broader context of cell biology in *Power, Sex, and Suicide: Mitochondria and the Meaning of Life* (2005). When we had our helpful conversations with Frank Harold in 2004, it was our understanding then that Frank had only recently been engaged in discussions of chemiosmosis with Nick Lane—no doubt our own conversations benefitted from those interactions. Along with having parallel conversations, Nick Lane also acknowledges Frank Harold's contributions, and the specific importance, as in our chapter, of Harold's 1986 and 2001 works. We acknowledge these common sets of roots and note that Lane's book was useful to us in its own right, and commend it to our readers.

The genesis of this chapter is credited to Norman Pace, who contributed to our first conference, in 2003, his understanding of the work of chemiosmosis as well as the role of such work in the early origins of life. Norm's belief that a fundamental biological work process could be the foundation for understanding work in the way this volume presents came to replace any number of other proposed concepts for the foundation of this exploration and placed it on firm footing. Throughout this process, Norm also contributed his time, creative thought, and his support, unstintingly and however asked.

References

Alberts, B., Johnson, A., Lewis, J., Raff, M., Roberts, K., & Walter, P. (2008). *Molecular biology of the cell* (5th ed.). New York: Garland.

Atkins, P. W. (2007). *Four laws that drive the universe*. Oxford: Oxford University Press.

Atkinson, D. E. (1965). Biological feedback control at the molecular level. *Science, 150*(3698), 851–857.

Baymann, F., Lebrun, E., Brugna, M., Schoepp-Cothenet, B., Giudici-Orticoni, M. T., & Nitschke, W. (2002). The redox protein construction kit: Pre-last universal common ancestor evolution of energy-conserving enzymes. *Philosophical Transactions of the Royal Society of London. Series B, Biological Sciences, 358*(1429), 267–274.

Beis, I., & Newsholme, E. A. (1975). The contents of adenine nucleotides, phosphagens and some glycolytic intermediates in resting muscles from vertebrates and invertebrates. *Biochemical Journal, 152*, 23–32.

Boutilier, R. G. (2001). Mechanisms of cell survival in hypoxia and hypothermia. *Journal of Experimental Biology, 204*, 3171–3181.

Boyer, P. D. (1997). The ATP synthase—A splendid molecular machine. *Annual Review of Biochemistry*, *66*, 717–749.

Bray, D. (2002). Bacterial chemotaxis and the question of gain. *Proceedings of the National Academy of Sciences of the United States of America*, *99*(1), 7–9.

Brooks, D. R., & Wiley, E. O. (1988). *Evolution as entropy: Toward a unified theory of biology* (2nd ed.). Chicago: University of Chicago Press.

Chapman, A. G., Fall, L., & Atkinson, D. E. (1971). Adenylate energy charge in *Escherichia coli* during growth and starvation. *Journal of Bacteriology*, *108*(3), 1072–1086.

Falke, J. J., Bass, R. B., Butler, S. L., Chervitz, S. A., & Danielson, M. A. (1997). The two-component signaling pathway of bacterial chemotaxis: A molecular view of signal transduction by receptors, kinases, and adaptation enzymes. *Annual Review of Cell and Developmental Biology*, *13*(1), 457–512.

Gogarten, J. P., Kibak, H., Dittrich, P., Taiz, L., Bowman, E. J., Bowman, B. J., ... Oshima, T. (1989). Evolution of the vacuolar H^+-ATPase: Implications for the origin of eukaryotes. *Proceedings of the National Academy of Sciences of the United States of America*, *86*(17), 6661–6665.

Gogarten, J. P., Starke, T., Kibak, H., Fishman, J., & Taiz, L. (1992). Evolution and isoforms of V-ATPase subunits. *Journal of Experimental Biology*, *172*(1), 137–147.

Hardie, D. G., & Hawley, S. A. (2001). AMP-activated protein kinase: The energy charge hypothesis revisited. *BioEssays*, *23*(12), 1112–1119.

Harold, F. M. (1974). Chemiosmotic interpretation of active-transport in bacteria. *Annals of the New York Academy of Sciences*, *227*, 297–311.

Harold, F. M. (1977). Ion currents and physiological functions in microorganisms. *Annual Review of Microbiology*, *31*, 181–203.

Harold, F. M. (1986). *The vital force: A study of bioenergetics*. New York: W.H. Freeman.

Harold, F. M. (2001). Gleanings of a chemiosmotic eye. *BioEssays*, *23*(9), 848–855.

Hill, J. W., & Kolb, D. K. (2007). *Chemistry for changing times* (11th ed.). Upper Saddle River, NJ: Pearson Prentice Hall.

Hochachka, P. W., & Somero, G. N. (2002). *Biochemical adaptation: Mechanism and process in physiological evolution*. New York: Oxford University Press.

Lane, N. (2005). *Power, sex, suicide: Mitochondria and the meaning of life*. New York: Oxford University Press.

Leslie, A. G. W., & Walker, J. E. (2000). Structural model of F_1-ATPase and the implications for rotary catalysis. *Philosophical Transactions of the Royal Society of London. Series B, Biological Sciences*, *355*(1396), 465–471.

Levin, R. A. (1984). Physiology and cellular biochemistry of hypothermia: A biochemical model for accidental hypothermia in humans. *1984 Olympic Scientific Congress Abstracts, Environmental Factors and Sport*, 3–4.

Manson, M. D., Tedesco, P., Berg, H. C., Harold, F. M., & Vanderdrift, C. (1977). Protonmotive force drives bacterial flagella. *Proceedings of the National Academy of Sciences of the United States of America, 74*(7), 3060–3064.

Mitchell, P. (1979). Keilin's respiratory chain concept and its chemiosmotic consequences. *Science, 206*, 1148–1159.

Nicholls, D. G., & Ferguson, S. J. (1992). *Bioenergetics 2*. London: Academic Press.

Nicholls, D. G., & Ferguson, S. J. (2002). *Bioenergetics 3*. London: Academic Press.

Nisbet, E. G., & Sleep, N. H. (2001). The habitat and nature of early life. *Nature, 409*(6823), 1083–1091.

Odling-Smee, F. J., Laland, K. N., & Feldman, M. W. (2003). *Niche construction: The neglected process in evolution*. Princeton, NJ: Princeton University Press.

Pace, N. R. (1997). A molecular view of microbial diversity and the biosphere. *Science, 276*, 734–740.

Pace, N. R. (2001). The universal nature of biochemistry. *Proceedings of the National Academy of Sciences of the United States of America, 98*(3), 805–808.

Rabinbach, A. (1990). *The human motor: Energy, fatigue, and the origins of modernity*. New York: Basic Books.

Raison, J. K. (1973). The influence of temperature-induced phase changes on the kinetics of respiratory and other membrane-associated enzyme systems. *Journal of Bioenergetics and Biomembranes, 4*(1), 285–309.

Robertson, C. E., Harris, J. K., Spear, J. R., & Pace, N. R. (2005). Phylogenetic diversity and ecology of environmental Archaea. *Current Opinion in Microbiology, 8*(6), 638–642.

Schopf, J. W. (1999). *Cradle of life: The discovery of Earth's earliest fossils*. Princeton, NJ: Princeton University Press.

Spear, J. R., Walker, J. J., McCollom, T. M., & Pace, N. R. (2005). Hydrogen and bioenergetics in the Yellowstone geothermal ecosystem. *Proceedings of the National Academy of Sciences of the United States of America, 102*, 2555–2560.

Stock, D., Gibbons, C., Arechaga, I., Leslie, A. G. W., & Walker, J. E. (2000). The rotary mechanism sf ATP synthase. *Current Opinion in Structural Biology, 10*(6), 672–679.

Stock, D., Leslie, A. G. W., & Walker, J. E. (1999). Molecular architecture of the rotary motor in ATP synthase. *Science, 286*(5445), 1700–1705.

Thompson, W. G., Foster, R. C., Eide, D. S., & Levine, J. A. (2008). Feasibility of a walking work-station to increase daily walking. *British Journal of Sports Medicine, 42*(3), 225–228.

Thurston, J. T., Burlington, R. F., & Meininger, G. A. (1978). Effects of low temperatures on rat myocardial Mg-ATPase and NaK-ATPase. *Cryobiology, 15,* 312–316.

Voet, D., & Voet, J. G. (2004). *Biochemistry* (3rd ed.). Hoboken, NJ: Wiley.

Walker, J. (2009). ATP synthase. Retrieved from http://www.mrc-mbu.cam.ac.uk/research/atp-synthase.

Woese, C. R. (1998). The universal ancestor. *Proceedings of the National Academy of Sciences of the United States of America, 95*(12), 6854–6859.

Woese, C. R., & Fox, G. E. (1977). Phylogenetic structure of the prokaryotic domain: the primary kingdoms. *Proceedings of the National Academy of Sciences of the United States of America, 74*(11), 5088–5090.

Simon Laughlin has pioneered the exploration of relationships between energy and information in living systems. It is easy for modern humans to idealize information as something that is free, including energy-cost free. For humans, though, that "free" information arrives over high fiscal- and energy-cost transmission lines to high fiscal- and energy-cost computers powered by extremely high fiscal- and energy-cost power stations. Laughlin's research in neurobiology conducts a similar analysis of the costs of moving information through biological systems and the energetic and evolutionary implications of these costs. His experimental measurements of the energy cost of information in insect compound eyes and his theoretical analysis of the energy required for neural processing in the mammalian brain show that neurons must, of necessity, use considerable quantities of metabolic energy to process information. These energy demands limit an organism's use of information. Brains have responded by evolving energy-efficient structures and mechanisms. If, after the discussion in chapter 1, work seems like it may require energy for lowly brutes and earlier ages of humans, but not for information-age sophisticates like us, Laughlin's account of energy-information trade-offs encourages a reassessment of this tempting conclusion.

2 Energy, Information, and the Work of the Brain

Simon B. Laughlin

Introduction

This chapter continues the theme *work meets life* by considering the energy used by nervous systems. Nervous systems process information using basic mechanisms of intracellular and intercellular communication that are elaborated to transmit and process signals in rapid, reliable, and far-reaching networks. A nervous system's primary function is to direct work. The work a nervous system directs is mainly in the form of muscular activity, but it extends to a range of activities, such as secretion and changes in body structure (e.g., sexual maturation), and includes learning and memory.

This chapter complements the widely accepted principle that nervous systems use information to direct energy into work with a second principle: Nervous systems use energy to direct information. Nervous systems must, of physical necessity, use considerable quantities of energy to channel and drive the flow of information. This is because information is carried and processed by electrical signals flowing through neural circuits, which, like the circuits in a mobile phone, consume energy. The quantities of energy required to drive electrical signals through neural circuits are large enough to constrain a nervous system's processing power. Therefore, energy demands influence the function and design of brains.

We start by pointing out that the two relationships, *information directs energy,* and *energy directs information,* are two sides of the same coin, for information and energy are, along with materials, the ingredients of life. This chapter defines *information,* describes how information is quantified as bits, and relates these bits to the ability of a nervous system to determine the state of an animal's surroundings and to generate appropriate action. Building on this foundation, measurements of the relationship between energy and information in nervous systems are described from experiments made in fly visual systems, which show that the neural coding of information is, by necessity, an energy-intensive process. Furthermore, energy demands are inflated by two behavioral requirements: speed and accuracy. We demonstrate that the quantities of energy used for neural processing are sufficient to limit processing power, by developing energy budgets for mammalian cerebral cortex. This combination of information theory, measurement of energy and information in nervous systems, and development of energy budgets for brains leads to the conclusion that the need to be energy efficient has left its mark on the architecture and function of the human brain. Consequently, the energy that a brain requires to process information influences the brain's ability to direct work.

The Interplay between Information and Energy in Biology

The Roles of Genetic and Extragenetic Information

The fact that information is an integral part of life is well known (Carroll, 2005). Genes transmit information from one generation to the next, so that like begets like. Molecular biologists have been spectacularly successful in discovering how this information is coded in DNA and RNA. We can read the genetic code—for example, the human genome—and we know how cells translate this code into the proteins that make cells work. It is less well known that the genetic information that is read out from genes spawns a much larger body of information. This information lives in the extensive signaling networks that cells construct, largely from proteins. These extragenetic networks control and coordinate the construction and operation of the complicated vehicle that propels genes to the next generation: the organism.

Extragenetic networks use a variety of mechanisms to exchange information for a multitude of purposes. The mechanisms operate over distances that range from the atomic to the global. As examples: Molecules interact directly at the atomic level when a hormone is bound to a receptor; messenger molecules and transcription factors move through a cell's nucleus and cytoplasm to coordinate many of the cell's operations; chemical, mechanical, and electrical signals, such as hormones, growth factors, and neural signals, travel through tissues to control development and function; and animals communicate with chemical, electrical, and mechanical signals—scenting, buzzing, and singing to their mates. The long-distance record for communication between animals belongs to whales, whose mechanical signals—whale songs—cross oceans (Dusenbery, 1992).

Most of these extragenetic networks share a common purpose: directing cells, organs, and organisms to gather and apply the materials and the energy that are needed for survival and reproduction. In other words, the information flowing through these networks is directing work, including the chemical work of gene replication and protein synthesis; the coordination of the chemical, osmotic, and mechanical work required to build and maintain cells; and the coordination of cells that is necessary for organs and organisms to function. Thus, the information that living systems collect, process, and transmit to direct their work is an essential ingredient of life.

A Thermodynamic Perspective on Information's Role in Living Systems

It is worth stepping back to consider the physics that links information to work in living systems. Information and work are essential ingredients in life's struggle to impose order on a disordered universe (Schrödinger, 1944). From a thermodynamic perspective, the atoms of our grand system, the known universe, can be arranged in many combinations, each defining a state. The state of the universe is constantly changing, subject to the laws of thermodynamics. The first law of thermodynamics says that, as the state changes, energy is conserved; the second law says that disorder (entropy) increases so that the universe as a whole becomes more random, more uncertain, and less predictable over time. Against this backdrop of inevitable disorder, living organisms stand out as highly improbable systems.

Organisms are improbable in the sense that they are unlikely to form spontaneously. You would have to shake up the atoms in the universe for an implausibly long time for them to spontaneously coalesce into an amoeba or even a tiny part of a bacterium, such as its flagellum. The probability that, if you carried on shaking, you would get a second amoeba a few hours later beggars belief. How then do living organisms make the incredibly unlikely happen, time after time? Organisms use energy from their surroundings to adopt and maintain their implausible states: They construct order. To self-replicate, organisms depend on information—both the information within genes and the information in the network spawned by genes—to specify structure and action. This information directs work, from genes to make proteins and in the extensive extragenetic networks that make and maintain cells. Because work is directed energy, information and energy are, along with materials, the ingredients of life. But what, precisely, is *information*?

Defining and Measuring Information

Defining the Unit of Information: The Bit

Information improves our knowledge of what is happening and what to do. *Information theory* quantifies the improvement of knowledge and the depth of instruction by going to the heart of these matters—the reduction of uncertainty (Pierce, 1962; Shannon & Weaver, 1949). Uncertainty is reduced by specifying which members of a set of alternatives are more likely; information theory defines the unit of information, the bit, in terms of the simplest elementary choice. The bit is the quantity of information required to decide between two equally likely alternatives.

Bits are readily combined to cover large numbers of possibilities and eventualities. A small number of choices, and, hence, bits, will decide among a much larger number of alternatives. For example, in a game of 20 questions, the successful player runs through a list of binary alternatives (a decision tree) that converges upon the target. The more complicated the system, the bigger the decision tree and the greater the number of bits needed to navigate its branches, but, in the end, causal relationships within the system can be boiled down to bits. Indeed, over the past 30 years, computers have used increasingly complicated patterns of bits to describe more and more of our surroundings, our decisions, and our actions.

Calculating Numbers of Bits

The number of bits required to specify a given situation or outcome is calculated from the likelihoods of each of the alternatives (i.e., the probabilities of each possible situation or outcome), as demonstrated with the example in figure 2.1. In the first condi-

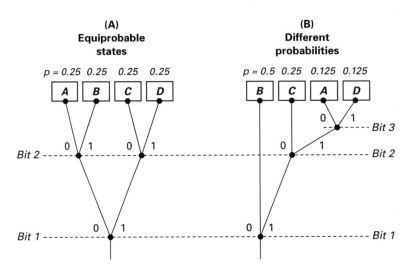

Figure 2.1

Basic approaches to information in bits. The decision tree shows how two successive binary choices specify the state of a four-state system when (A) all states are equally likely and (B) some states are more likely than others. Note that, because the choices are between equally likely alternatives, they all involve one bit of information.

tion in this example, a system exists in one of four equiprobable states: *A, B, C,* or *D.* All states are equally likely; therefore, the probability of finding the system in state *A, p(A),* is equal to *p(B), p(C),* and *p(D).* Because the system must always be in one of these four states, these probabilities add up to 1.0, giving

$$p(A) = p(B) = p(C) = p(D) = 0.25.$$

The state of this system can be specified by making two successive binary choices (figure 2.1A, equiprobable states). The first choice decides between equally likely pairs of states, *(A,B)* and *(C,D).* The second choice decides between the two equally likely members of the pair identified by the first choice. Because these choices are between two equally probable alternatives, each involves exactly one bit of information. We find, therefore, that two bits of information identify which of four equally likely alternatives is correct. The choices form a decision tree (figure 2.1), and the bits provide a binary code for states

$$A = 00, B = 01, C = 10, D = 11.$$

Note that the number of bits, two, is the log to the base 2 of the number of alternatives, four. In general, the number of bits of information, *I,* that is required to specify a system that can be in any one of *U* equally likely states is given by

$$I = \log_2(U) \text{ bits.} \tag{2.1}$$

Thus, an 8-state system is specified by three bits, a 16-state system by four bits, and so on. Note that because uncertainty increases with the number of possible states, the number of bits required to remove this uncertainty also increases. Thus, as defined by information theory, bits reduce uncertainty.

Although the account so far sounds like basic computer science, information theory is not restricted to digital systems. The theory was not constructed around digital computers; it was developed to handle the large numbers of unequal likelihoods found in everyday life (Pierce, 1962; Shannon & Weaver, 1949). Information theory can use the probability distribution of system states to calculate the number of bits required to specify which state the system is in. To demonstrate this versatility, in the second condition in figure 2.1, we set up our four-state system with an asymmetric probability distribution

$$p(A) = 0.125, p(B) = 0.5, p(C) = 0.25, p(D) = 0.125.$$

Given these probabilities, the resulting decision tree (figure 2.1B, different probabilities) specifies *A, B, C,* and *D* using binary choices between equally likely alternatives. Initially, this scheme appears to be inefficient, because it involves three bits when, as earlier shown in the equiprobable states condition in figure 2.1A, a two-bit code (*A* = 00, *B* = 01, *C* = 10, *D* = 11) is able to specify the four states, no matter what their probabilities are.

However, when we repeatedly use the new decision tree to determine the system's state on many different occasions, we will average less than two bits per determination. The full three bits are used relatively infrequently because they specify the two least probable states A and D. Figure 2.1B shows that three bits are used on 25% of our climbs through the decision tree. Another 25% of our climbs use two bits to reach C, and 50% of the climbs go straight to B using just one bit. Thus, on average, we use:

0.25 × 3 bits + 0.25 × 2 bits + 0.5 × 1 bit = 1.75 bits per determination of state

Note, again, how bits relate to uncertainty. It took one bit to reach the most likely state, the one that occurs with the greatest certainty, B, and three bits to get to the least likely states, A and D. In general, if the probability of occurrence of a state u is p(u), the number of bits of information required to specify that state, I_U, is

$$I_U = \log_2(1/p(u)) = -\log_2(p(u)) \tag{2.2}$$

which, when states occur equally often, becomes equation (2.1), because when there are U equally likely states, the probability of encountering any one of them is $1/U$. Furthermore, the average number of bits required to specify a system over many independent determinations also decreases with decreasing uncertainty. Knowing that the system is in state B for 50% of the time, and is either in state B or state C for 75% of the time, reduces the requirement from 2 bits when all states are equally likely to 1.75 bits.

The simple four-state system also demonstrates that bits of information reduce uncertainty without necessarily eliminating uncertainty. In the equiprobable case in figure 2.1A, a single bit reduces uncertainty from one out of four to one out of two. Thus, if you do not have sufficient bits, you cannot get the right answer, but every bit helps to decrease your uncertainty.

A simple system helps us to understand how bits reduce uncertainty, but how might we calculate the number of bits involved in complicated systems, with many unequal alternatives? Fortunately, we are spared the labor of having to construct ever-more-complicated decision trees. The results yielded by custom-built trees are generated by a master equation that Shannon derived rigorously from first principles (Shannon & Weaver, 1949) to handle complicated situations, in dealing with texts, telephones, and telegraphy. If a system occupies a set of x states, and each state occurs with a particular probability p(x), then the set of x states represents H(x) bits of information where

$$H(x) = -p(x)\sum_1^x \log_2(p(x)). \tag{2.3}$$

Note that the results generated by this equation agree with common sense. If we already know the answer (e.g., p(A) = 1, p(B) = p(C) = p(D) = 0), then H(x) is zero, and no bits are required.

Shannon realized that this powerful equation can be used in two ways. Not only does the equation specify the amount of information required to describe a system, but it also calculates the amount of information that a system can communicate by switching between a set of signal states. For communication, x could be the set of signals used to transmit messages; for example, the letters of an alphabet, digital bytes traveling in a computer network, or a stream of action potentials being discharged by a neuron in the brain. In these cases, $p(x)$ describes the frequency with which each signal is used, and $H(x)$ is the number of bits communicated by the set of signals.

To summarize this elementary presentation of information theory: Information is required to specify the state of a system. In other words, information increases confidence about what is there and what will happen. Consequently, information is quantified in terms of its ability to reduce uncertainty; the more uncertain one is, the more information one requires to set the world to rights. The basic unit of information, the bit, is digital; it is the information required to resolve the choice between two equally likely possibilities. However, just as horsepower does not require you to feed your car oats, bits are not restricted to digital systems; they are universal. Indeed, physicist John Wheeler's proposition that "everything is information" because "no phenomenon is a phenomenon until it is a recorded phenomenon" (Misner, Thorne, & Zurek, 2009, pp. 44–45) is being avidly pursued by physicists, cosmologists, and computer scientists (Bekenstein, 2003; Misner, et al., 2009). Furthermore, because we live in a universe in which physical laws attribute effects to causes, information is all around us (Leff & Rex, 2003; Lloyd, 2000).

Humans put this all-pervasive quantity, information, to many uses, because the specification of states is fundamental to human existence. To give an example from business, sets of states include the list of share prices in New York, the pattern of bytes that transmits these prices via a computer link to London, and the broker's instructions to buy, sell, or pass, attached to a list of shares. To give an example of the brain's role in specifying behavior, information is in the pattern of light that your eye records to monitor the state of the world ahead. The brain then processes this information to interpret the state of the way ahead and to formulate the necessary instructions about what to do. This information is sent out as a pattern of commands to the muscles that, for instance, when driving a car, prompt a driver to turn the steering wheel to a new state, a state that is intended to increase the likelihood that the driver will survive. In these examples, information is useful because, by decreasing uncertainty, it has the potential to increase the likelihood of favorable outcomes. But is it practical to collect, use, and distribute the huge quantities of information that surround us?

The answer to the question—Is information processing practically boundless?—is no. Google, the world's single most intensive collector, processor, and distributor of information, will soon hit a limit of capacity. Industry leaders have told the *Guardian* newspaper that Google faces a "perfect storm." Google is in a tight situation. On the one side, it is committed to satisfy customers' demands for information, and these demands are rising exponentially. On the other side, the company faces steep increases in energy costs. Google's information processors, a giant bank of servers housed in a complex near The Dalles, Oregon, are forecast to use as much electrical power by the year 2011 as would a city of over a quarter of a million people (Johnson, 2009): There is no escaping the fact that information requires energy.

Information and Energy

Information involves energy because something must happen to increase the certainty of a situation; for example, a measurement, a movement, or a command. This involvement is apparent in a communication system (figure 2.2), where a source emits a set of signals, s_1, s_2,... s_n, that travel though a channel and are picked up by a receiver. The signals will carry information from the source to the receiver if they increase the receiver's certainty about the state of the source. For this to happen, a change in the state of the source must produce a distinct change in the state of the signal that can be discriminated by the receiver; the mechanisms implementing this chain of reliable communication are driven by energy.

Theoretical studies of the thermodynamics of communication and practical studies of communications systems demonstrate that a bit requires a minimum amount of energy (Landauer, 1996; Lloyd, 2000; Pierce, 1962). This thermodynamic lower limit is around kT joules per bit, where k is the Boltzmann constant, and T is the absolute temperature in degrees Kelvin. The amount is so small, approximately 3×10^{-21} J at room temperature, that the potential energy released by dropping a pen from desk to floor could support roughly 10^{19} bits. However, powerful practical systems, such as computers and brains, operate far above this thermodynamic lower limit. In 2009, the world's most powerful supercomputer, the IBM Roadrunner, uses 2.4 megawatts (MW) to process 10^{15} bits per second. Thus, the IBM Roadrunner's energy expenditure of 5×10^{-8} J per bit is 10^{13} times greater than the thermodynamic minimum. The human cerebral cortex consumes less than 10 W to carry out approximately 10^{15} synaptic

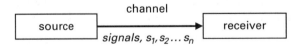

Figure 2.2
Shannon's basic communication system. A source sends a series of signals, s_1, s_2, ... s_n, via a channel to the receiver. The quantity of information transmitted by the signals depends upon the degree to which these signals reduce the receiver's uncertainty about the state of the source. For commands, it is the degree to which the signals specify the state of the receiver.

operations per second and, if one guesses that one synaptic operation is equivalent to processing at least one bit, the cortex uses 10^{-14} J per bit. Although the cortex's energy expenditure per bit makes it one million times more energy efficient than the IBM Roadrunner computer, the cortex still operates 100,000 times above the thermodynamic minimum. Why do these powerful devices consume so much energy to handle their bits?

Computers and nervous systems are forced to use a great deal of extra energy because they have to confine, direct, and deliver signals to specific destinations and process them. These costs generally increase with the rate at which information is handled. Theoretical studies of energy limits to computation (Leff & Rex, 2003; Lloyd, 2000) find that the energy needed to perform a set of computations increases with the speed with which they are performed. The rates at which high-performance mobile phones and computers run down their batteries seem to support the suggestion that speed increases costs, but does this relationship between speed and energy prevail in nervous systems? To answer this question, we must measure the cost of neural information, in energy per bit. To date, it has only been possible to make this measurement in one set of neural systems: flies' compound eyes.

The Energy Cost of a Neural Bit

Fly photoreceptors can be used to establish the energy cost of a bit of information, because they satisfy two requirements. First, we have an unusually complete understanding of the information that they code, and, second, we have a detailed description of the mechanisms that code this information. This section describes the coding mechanisms and the amount of information coded, and then, by determining photoreceptor energy usage, calculates how much energy a cell uses to handle one bit of information.

How Fly Photoreceptors Code Pictorial Information

The fly's compound eye forms a mosaic image of its visual surroundings in which each facet of the compound eye is equivalent to a single pixel (figure 2.3). The light level in each pixel is registered by a group of eight photoreceptors. Each photoreceptor absorbs light to produce an analog electrical signal whose amplitude goes up and down with the light level (figure 2.4).

This electrical signal is generated by tiny molecular pores, ion channels, that allow electrically charged ions to cross the cell membrane. Potassium channels allow positively charged potassium ions to leave the cell, and light-gated channels allow positively charged sodium ions to enter (figure 2.5). The voltage across the photoreceptor cell membrane, the membrane potential, depends on the balance between positively charged potassium ions leaving the cell to make the voltage more negative

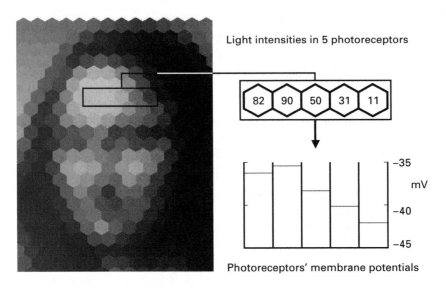

Figure 2.3
The fly's eye view of a well-known picture. The visual image is composed of a hexagonal array of pixels, one for each facet of the compound eye. The row of five cells illustrates that photoreceptors capture the light intensity in each pixel and convert this into an electrical response for transmission to the brain. Note that this diagram is a snapshot—a single frame in a long movie. Pixilated image courtesy of Brian Burton.

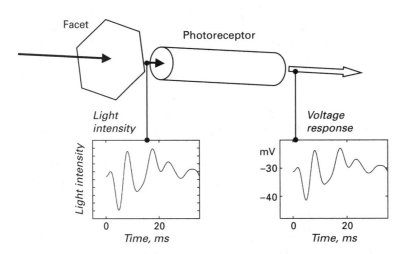

Figure 2.4
Photoreceptors code the ever-changing pattern of light in a pixel as an analog voltage signal. The light entering a single facet of the compound eye changes intensity continuously as the eye moves relative to the visual scene. The photoreceptor converts this continuously changing optical signal into an electrical signal that tracks the light intensity.

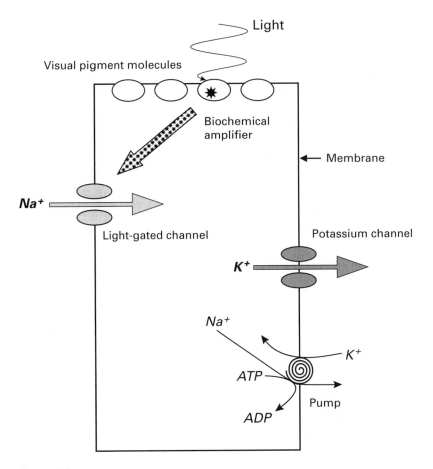

Figure 2.5
A simplified diagram of the mechanism used by fly photoreceptors to convert light into an electrical signal.

and positively charged sodium ions entering through light-gated channels to make the voltage more positive.

In the dark, most of the light-gated channels are closed, potassium ions flowing through potassium channels dominate, and they push the membrane potential down to −60 mV. When a photon of light is absorbed by a molecule of the visual pigment in the photoreceptor membrane, the molecule activates a biochemical amplifier (figure 2.5). The amplifier opens light-gated channels, and this allows sodium ions to enter the cell and drive the membrane to a more positive value. The amplitude of this upward swing increases with light level because the number of channels opened in response to light increases with the rate at which photons are being absorbed by visual pigment molecules. Conversely, when the light level falls, the number of channels opened by light goes down, the balance tips back in favor of potassium, and the membrane potential drops to a more negative value. This coupling of the opening and closing of membrane ion channels to the absorption of light ensures that the photoreceptor membrane potential continuously tracks the light level.

The photoreceptor's membrane potential carries information because it is coupled to light level by a well-behaved, deterministic mechanism (figures 2.4 and 2.5). Uncertainty about the state of a pixel is reduced because one can infer its light level from the membrane potential. This voltage signal is transmitted passively to the photoreceptor's output terminal in the first visual ganglion, where its information is delivered to visual neurons via synapses. By processing the outputs from the array of photoreceptors, the fly's brain is able to compare the states of pixels across the retinal image, and this enables the visual system to identify a larger set of states relating to the animal's surroundings. One of these could be the last set of states a fly sees: the expanding patch of dark pixels caused by the rapid advance of a flyswatter. It follows that an increase in the rate at which its photoreceptors code information could, by increasing the fly's certainty about the state of its surroundings, help it to spot the swat sooner. As described in the next section, theory shows that photoreceptors must respond quickly and accurately to achieve high information rates.

How Speed and Precision Determine Information Rate

The photoreceptor codes light level as a continuous voltage signal that goes up and down with light intensity (figure 2.4). This is an analog signal, similar to the electrical signals transmitted and received by a 1940s telephone. When Claude Shannon worked for Bell Telephone, he developed the theory that gives the rate at which this type of analog signal transmits information (Shannon, 1949). His derivation is elegant, sophisticated, powerful, and widely used by communication engineers. Shannon used Fourier analysis to convert the analog response into its frequency components and then summed the information carried at each frequency. However, one need not go into these details to see how speed and precision of response determine the rate at which an analog system transmits information (Pierce, 1962).

The following simplified account reveals the roles played by speed and accuracy in determining information rate. A continuously varying analog signal can be reduced to a series of rapid jumps between distinct voltage levels (figure 2.6). This is how music is recorded onto a CD. The waveform, continuously varying sound pressure, is measured 44,100 times per second, and each measurement is recorded as a 16-bit digital integer that assigns the sound pressure to one of 2^{16} (65,536) levels. The rate at which this process records information is the number of bits per sample multiplied by the number of samples per second. Of course, an analog signal does not jump between

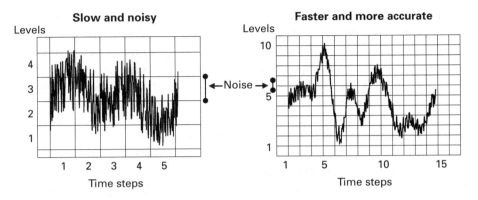

Figure 2.6
A slow and noisy analog response codes fewer time intervals and amplitude levels than a faster, more accurate response. Because it is using fewer signal levels and transmitting fewer signals per second, the slow, noisy cell transmits fewer bits per second.

discrete response levels at a fixed rate, but physical limits to the speed and accuracy of signaling make it behave as if it does.

Discrete levels are created by uncontrolled random fluctuations in response, called noise. For a change in the response to be reliably attributed to a change in the signal, it must exceed the noise. Thus, noise divides the total range of response into a series of discrete levels, each corresponding to one just noticeable difference (JND) in signal (figure 2.6). The total number of these JNDs is the total response range, signal, S, plus the noise, N, divided by the noise. (The term S/N in equations (2.4) and (2.5) below is referred to as the *signal-to-noise ratio* and is described later in the chapter.) As a result, the number of discrete signal levels, L, that are coded as an analog signal is given by

$$L = (S+N)/N = 1 + S/N. \tag{2.4}$$

From equation (2.1), the number of bits per level, I_L, is given by

$$I_L = \log_2(L) = \log_2(1 + S/N). \tag{2.5}$$

The maximum rate at which changes in signal level are reported is set by the speed with which the analog response is able to change amplitude (figure 2.6). Analog signals that change sluggishly report fewer changes per second than brisk responders. The ability of a signaling system to follow rapid changes is commonly determined by measuring the responses to signals that oscillate sinusoidally at various frequencies. This frequency response yields the bandwidth, B, which for a photoreceptor specifies the highest effective rate at which its signal switches levels. Thus, the information rate, I, is given by

$$I = B\,I_L = B\log_2(1+S/N). \tag{2.6}$$

This equation for analog information rate reflects common sense—brisk, accurate responses clearly convey more information than sluggish, unreliable ones (figure 2.6). The equation also reminds us that bits reduce uncertainty. In the presence of noise, the receiver is forced to ask the question, is this change in response produced by a change in the signal or by random changes due to noise? Consequently, as noise increases, uncertainty increases, and bits decrease.

The Bit Rate Achieved by a Fly Photoreceptor's Analog Signal

Knowing how analog signals convey bits of information, we return to our original question: How many bits do photoreceptors transmit? The answer, obtained by applying Shannon's elegant method to the responses recorded from blowfly photoreceptors, is 1,000 bits per second (de Ruyter van Steveninck & Laughlin, 1996). To derive the energy cost of these bits, we must estimate the amount of energy that is used to generate the photoreceptor response. The resulting value gives considerable insight into the reasons that neurons use significant quantities of energy to process information at reasonable rates.

The Energy Cost of Information in a Fly Photoreceptor

The energy cost is calculated by considering the mechanism that drives current across the photoreceptor membrane to produce the voltage response (figure 2.5). The current is carried by sodium ions and potassium ions, which are driven through channels by concentration gradients. Sodium ions flow in because the sodium concentration is higher outside the photoreceptor than inside, and potassium ions flow out because the potassium concentration is higher inside than outside. This system for driving currents across the membrane is equivalent to a pair of electrical batteries, with voltages that represent the driving forces for sodium ions and potassium ions, respectively. Because the driving force for an ion is provided by its concentration gradient, a battery's voltage will drop as concentration gradients fall. Therefore, signaling will run down these ionic batteries, as ions flow from the region of higher concentration to the region of lower concentration. To maintain the voltages on its ionic batteries, the photoreceptor must maintain the concentration gradients, and, to achieve this end, the photoreceptor uses metabolic energy to pump the ions back across the membrane from whence they came.

The pumps are molecules in the membrane (figure 2.5) that use the energy released by the hydrolysis of an adenosine triphosphate (ATP) molecule to extrude three sodium ions and import two potassium ions. (See chapter 1.) The pumps work vigorously to maintain the gradients by pumping back the same number of ions as flowed through channels. These ionic fluxes, the currents driving the light response, can be estimated from measurements of the photoreceptor membrane voltage and resistance. Knowing the gearing ratio of the pump, in ions per ATP, one can convert this ionic flux into the rate of energy consumption. When signaling in daylight, a fly photoreceptor must consume about 10 billion (10^{10}) ATP molecules per second to keep its ionic batteries charged (Laughlin, de Ruyter van Steveninck, & Anderson, 1998).

Now we can work out the energy cost of a bit of information. In daylight, the photoreceptor transmits 1,000 bits per second and consumes 10^{10} ATP molecules per second to keep its ionic batteries charged. Thus, one bit costs this neuron 10 million ATP molecules (Laughlin et al., 1998). Using the energy released by hydrolyzing a single ATP molecule to convert from biology's energy euro, the ATP molecule, to physically respectable joules, we find that a photoreceptor uses more than $10^8 \ kT$ J per bit. What insights can we gain from these abstract measures of energy per bits? What does this mean, how does it compare with consumption in other systems, and does it have any impact on the fly's total energy budget?

This measurement of the energy cost of information in a neuron prompts several observations. Our particular neuron, the fly photoreceptor, operates 100 million times above the thermodynamic lower limit of kT joules per bit. Moreover, we can guess why it is using so much energy. Like other neurons, fly photoreceptors have to do several things with their bits of information; namely, they receive, register, process, and transmit them. The thermodynamic limit applies to elementary events that register information, such as an atom or a molecule switching between states to signify a single bit. For the photoreceptor, this elementary event is the first step in phototransduction, the activation of a visual pigment molecule by a photon hit. But the photoreceptor has to go further. It processes information, by summing signals from multiple hits, and then transmits the result to its output terminal.

Both processing and transmission use energy and, because it is electrical currents that are summed, and voltages that are transmitted, this energy is provided by the photoreceptor's ionic batteries. Supplying this energy places fly photoreceptors high on the list of energy-demanding cells. These photoreceptors are high energy burners, with a power density (rate of energy consumption per gram of tissue) greater than human muscles working flat out on a 100-meter sprint. Fly photoreceptors maintain this high rate throughout the day, but sprinters cannot.

Provisioning its high-performance photoreceptors has quite an impact on the blowfly's energy budget. In daylight, the eyes' photoreceptors consume 8% of the fly's resting energy consumption. This investment in high rates of information uptake seems worthwhile, because the fly has to direct and control its vigorous flight maneuvers. In the next section, we will see that high rates make bits more expensive.

Information Cost and Information Capacity

A comparison between photoreceptors in different species (Niven, Anderson, & Laughlin, 2007) shows that the cost of a bit of information increases with the maximum

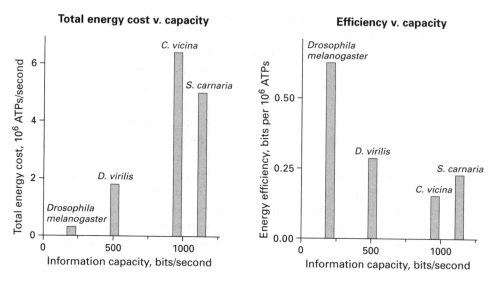

Figure 2.7
Energy costs increase and efficiency decreases as photoreceptor performance increases. Data are from four flies: the small fruit fly *Drosophila melanogaster*; the intermediate-sized *Drosophila virilis*; and two large flies, the blowfly *Calliphora vicina* and the flesh fly *Sarcophaga carnaria* (Niven et al., 2007).

rate at which a photoreceptor can transmit bits. The photoreceptors of the small fruit fly, *Drosophila melanogaster*, achieve maximum information rates of around 200 bits per second.

However, this fivefold drop in capacity is accompanied by a tenfold increase in energy efficiency, measured as bits per million ATPs (figure 2.7). In other words, cost increases with capacity, and this makes it more efficient to use low-rate neurons to transmit information. Measurements from two other species suggest a consistent trend (figure 2.7): Costs increase and efficiency falls as capacity rises. These relationships amount to a *law of diminishing returns*, as observed in cars where spectacular top speeds come with woeful fuel efficiencies, and the faster you want to go, the worse it gets (see chapter 3). As in cars, the photoreceptor's law of diminishing returns is determined by physics (Niven et al., 2007).

The Biophysics of Cost and Capacity
Fly photoreceptors exemplify two basic routes to increasing the amount of information coded by a neuron. This first route is increasing precision, which is done by increasing the signal-to-noise ratio, and, hence, the number of recognizably different response levels, L (equation (2.4); figure 2.6). The second route is to increase the speed of response and, hence, the bandwidth, B, to transmit more levels per second (figure

2.6). The photoreceptors of the two large flies, *Calliphora* and *Sarcophaga*, take both routes to achieve a higher information rate. The signal-to-noise ratio in the high-rate *Sarcophaga* photoreceptor is 50% higher than in the low-rate *Drosophila* photoreceptor, and the bandwidth is three times wider.

Both routes to improvement—precision and speed—involve using more energy. With the first route, photoreceptors achieve a higher signal-to-noise ratio by increasing the rate at which they record photon hits; this requires more light-gated channels and, hence, more current. The signal-to-noise ratio increases as the square root of the number of channels, but the energy cost increases more steeply. The cost is proportional to the ionic current, which increases with the number of channels. Consequently, the signal-to-noise ratio route to extra information is expensive and subject to the law of diminishing returns (Laughlin et al., 1998).

The second route, increasing speed to improve bandwidth, also demands extra energy. To signal more rapidly, neurons have to increase the flow of current across the membrane because they have to charge and discharge the capacitance of the membrane more quickly. (Readers familiar with biophysics will recollect that the membrane time constant is proportional to membrane resistance.) Blowfly photoreceptors lower their membrane resistance by inserting more potassium channels, and the resulting increase in ion flow has to be counteracted by pumping more ions, resulting in an increase in the consumption of ATP. Thus, experiments on fly photoreceptors show that the energy cost of information increases steeply with information capacity because, for biophysical reasons, precision and speed are expensive commodities.

These relationships between energy and information, first discovered in fly photoreceptors, apply widely in nervous systems (Laughlin, 2001). Consider the route of increasing capacity by improving signal-to-noise ratio. All neurons use ion channels to generate electrical signals but, because of random (thermodynamic) fluctuations, the channels operate unreliably. Consequently, as in fly photoreceptors, the signal-to-noise ratio tends to increase as the square root of the number of channels. Now consider the route of increasing response speed. Neurons generally charge and discharge their membrane capacitance to produce electrical signals, and this can be done quickly only by increasing the membrane current, making this an expensive strategy. Thus, because precision and speed are expensive commodities, neurons must use significant quantities of energy to handle large amounts of information quickly. The next section describes the quantities of energy required and their effects on a brain's ability to process information.

Energy Limits Brain Work

Fainting fits and the damaging effects of interruptions in blood supply (e.g., stroke) demonstrate that the human brain is critically dependent on energy. Although the adult human brain is less than 2% of body weight, it is responsible for 20% of resting energy metabolism, and, in children around six years of age, it is 60%. A substantial proportion of cerebral energy is used for neural signaling. Neuronal consumption increases with the rate at which neurons fire electrical impulses, and the increases are confined to areas of heightened neural activity (Clarke & Sokoloff, 1999; Sokoloff, 2004). Because neurons are constantly active in the vigilant brain, the specific metabolic rate of neural tissue (rate of energy consumption per gram), is relatively high, equaling that of heart muscle in the resting animal (Ames, 2000).

The mammalian brain's energy demands are satisfied by a finely tuned supply system. The principal sources of energy, glucose and oxygen for aerobic respiration, are supplied by blood flowing through an elaborate cerebral vasculature. Spectacular advances in brain scanning have shown that the blood flow is locally controlled to match local demands, and this regulatory system has enabled neuroscientists to map the workings of the human brain (Shulman & Rothman, 2004). Magnetic resonance imaging is used to scan the brain and localize changes in blood supply, with a resolution of approximately 0.5 mm. Armed with this advanced technique, cognitive neuroscientists have produced over 50,000 scientific papers in the last 15 years, associating local changes in vascular supply with the performance of a huge range of cognitive tasks.

These functional imaging studies are helping to build a fascinating picture of how the human brain engages in cognition. There is continuous background activity in all regions of the brain, both when awake and asleep, and this takes most of the brain's energy. Performing a particular cognitive task involves relatively small increases in vascular supply in a small subset of local areas. Thus, the entire brain seems to be constantly active, perhaps taking care of a myriad of routine activities, such as consolidating past experience and maintaining a high level of vigilance, and, when we engage in a particular task, a small proportion of regions work a little bit harder. Given that all regions could, in principle, work a little bit harder, there is an element of truth in the popular idea that we only use a fraction of our brainpower at any one time. However, as described further below, any attempt to recruit all of the brain's processing power would be thwarted by excessive demands on its energy supply.

The widespread use of functional imaging raised the questions (Shulman & Rothman, 2004), how much energy are neurons using to process information? and, what is this energy used for? The majority of functional imaging studies examine blood supply to neurons in the outer layer of cerebral cortex, the cortical gray matter. Because all of the synapses that connect neurons into circuits are in gray matter, this layer is where information is processed. The underlying layer, cortical white matter, contains the network of cables that connect circuits in different regions. By developing bottom-up *energy budgets* for gray matter it proved possible to assign energy usage to the mechanisms that operate in circuits (figure 2.8), such as the transmission of signals

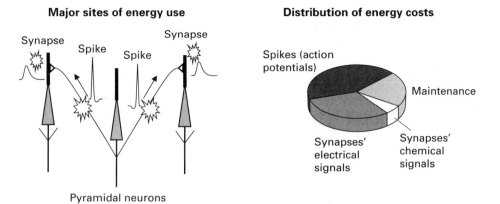

Major sites of energy use   **Distribution of energy costs**

Pyramidal neurons

Figure 2.8
The major sites of energy usage in the circuits formed by neurons in cortical gray matter, and the relative distribution of costs among these sites. Synapses' chemical signals are for the release and recycling of transmitter and vesicles, presynaptic calcium entry, and postsynaptic chemical (second messenger) responses. About 20% of energy is devoted to maintenance, including supporting glia cells. Values were calculated for rat neocortical gray matter (Attwell & Laughlin, 2001).

with action potentials, the generation of synaptic responses, and the recycling of chemical neurotransmitters (Attwell & Laughlin, 2001; Lennie, 2003).

The budgets show that most of gray matter's energy is used to drive fast electrical signaling by neurons (figure 2.8), and this pattern of usage has been confirmed in another region of the mammalian brain, the olfactory bulb (Nawroth, Greer, Chen, Laughlin, & Shepherd, 2007). As with fly photoreceptors, this energy is consumed by ion pumps to stop neurons' ionic batteries from running down. About 80% of neurons in gray matter are pyramidal cells. For this cortical neuron, the most demanding electrical signals are the action potentials that travel along its 4 cm of axonal connections within gray matter, and the postsynaptic potentials generated at its 8,000 synapses (figure 2.8). About 20% of the total energy is used to maintain the integrity of all of the cells in gray matter; for example, to generate resting potentials and turn over large molecules, especially proteins and lipids.

The bottom-up energy budget revealed why cortical gray matter is one of the most energy-demanding parts of the mammalian brain: Its neurons are densely packed and extensively interconnected. In the best documented gray matter, that of the mouse (Braitenberg & Schütz, 1998), there are 100,000 pyramidal neurons per cubic millimeter. Each neuron's 8,000 synaptic connections in gray matter are strung out along axon collaterals that are thin (0.3 micrometer (μm) diameter) and densely packed to achieve a wiring density of 4 kilometers (km) per cubic millimeter (mm^3).

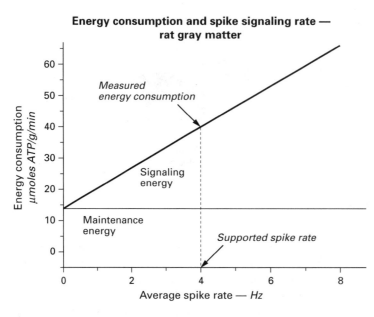

Figure 2.9
Energy consumption in cortical gray matter rises steeply with the level of signal traffic. Energy usage is given as the rate of ATP consumption per gram of tissue, and signal traffic is the mean rate at which cortical neurons transmit spikes to synapses (figure 2.8). This mean is the average across all neurons. Experimental measurements of energy usage (Sokoloff, 2004) suggest that signal traffic is limited to approximately 4 Hz.

The budget shows that gray matter energy consumption rises steeply with the rate at which its neurons signal with action potentials. This consumption rises from a baseline set by the fixed costs of maintenance (figure 2.9). The energy used per action potential is so large that the specific metabolic rate of gray matter can only sustain four signals per neuron per second. In other words, energy consumption severely limits the volume of signal traffic in cerebral cortex. Electrical signaling in the network of axons, synapses, and dendrites is using most of this energy. Less than 10% of energy consumed goes to chemical signaling within neurons, and this emphasizes that chemical communication is generally cheaper than electrical communication. However, electrical communication has the advantage of computing and transmitting information quickly over longer distances. Nervous systems can increase their energy efficiency by resorting to cheaper chemical means whenever possible. Many of the signals used for short-range information processing at synapses and within neurons are chemical and so are slow processes, such as the long-term storage of information and slow adjustments in excitability. Finally, less than 20% of energy is used to pay the fixed costs of maintaining the structure and integrity of neurons and their supporting glial cells. Cortical energy costs are dominated by the electrical signals that allow neurons to operate quickly over long distances. Once again, it is clear that speed is an expensive commodity.

Speed is not the only factor. Cortical processing is, of necessity, energy intensive—it comes with the job. Cortical neurons form the complicated networks that are responsible for making the widespread associations that support higher cognition, such as sensory perception, learning, formulating goals, making choices, planning actions, and predicting outcomes. A pyramidal neuron is well equipped to make wider associations because its collaterals form synapses that contact several hundred other pyramidal cells scattered over a considerable area. Because it is these synapses and collaterals that are using most of its energy (figure 2.8), gray matter is paying a high price for the ability to detect diverse patterns of cause and effect. This simple observation suggests that far-reaching powers of deduction are, by their nature, energy demanding, because, for the brain to search out new relationships, it must constantly correlate information from diverse sources. Thus, the inescapable relationships between information and energy in neurons provide good functional reasons to label primate cerebral cortex as "expensive tissue," whose energy demands have shaped its evolution (Aiello, Bates, & Joffe, 2001; Allman, 1999).

The Energy-Efficient Brain

If demands for energy place stringent limits on signal traffic, then nervous systems should have evolved mechanisms and structures that are energy efficient (Attwell & Laughlin, 2001; Laughlin, 2001). Our energy budget suggests how efficiency might be achieved by identifying the hot spots (figure 2.8) where savings can be made.

Saving on Spikes

The most obvious saving is to minimize the number of spikes (action potentials) used to execute a given task. This economy can be made by a number of means, all commonly used in nervous systems and particularly prominent in cortex. First and foremost, nervous systems can reduce the need to transmit signals by eliminating redundancy, which is defined by information theory as a response that carries no extra information because it can be predicted from existing signals (Shannon & Weaver, 1949). Many neurons, including cortical cells, eliminate redundancy by firing transiently when conditions change, and this is why our perception of the visual image fades within 3 seconds if the image is held completely stationary on the retina.

The cortex also saves on action potentials by using energy-efficient neural codes. These codes tend to maximize the quantity of information coded (i.e., representational capacity) within the constraint of a limited number of action potentials. One way to achieve efficiency is to distribute a small number of action potentials sparsely across a much larger group of neurons. To illustrate the efficiency of a sparse distributed code, imagine that a region of the brain has to represent 20 equally likely states of the world. These states could be coded by the number of action potentials fired by a single neuron in a given time interval; that is, 1 action potential for state 1; 2 action potentials for state 2, and so on. Because all 20 states occur equally often, this single neuron code will use an average of 10 action potentials per coding interval. Alternatively, one could use 20 neurons and specify the state as the identity of the particular neuron that fires an action potential in the coding interval. Because only one neuron in 20 is firing in a coding interval, this sparse code appears to be 10 times more efficient than the single-neuron code. However, the efficiency of a sparse code is strongly dependent on the fixed cost of maintaining the neurons that remain silent. If neurons cost a lot to maintain and little to fire, one should use a small population of neurons and fire all of them frequently (Levy & Baxter, 1996). The energy budgets suggest that the fixed costs of cortical neurons are relatively low—equivalent to little less than one action potential per second—and this makes sparse codes energy efficient (Attwell & Laughlin, 2001; Levy & Baxter, 1996). Indeed, to live within the cortical energy budget, monkey (and, by implication, human) neurons should fire, on average, less than once per second (Lennie, 2003). Sparse codes are found in visual cortex, where each neuron responds selectively to particular features, such as an edge of a particular length and orientation at a particular position in the visual field. Because a given neuron's special feature is seldom present in its little patch of the image, the neurons seldom signal. Thus, selectivity for features generates a sparse code.

Saving on Synapses

Significant savings can be made by reducing the use of synapses, because these structures take over one third of the cortical energy budget. Only 25% of a rat pyramidal neuron's 8,000 synapses are used at any time, because synapses are depressed according to previous patterns of neural activity. The depression and upregulation of synapses provide a substrate for learning patterns by selectively strengthening connections between cells whose activities coincide (Hebb, 1949). When depression is effective enough to turn down a large proportion of synapses, then large quantities of energy are saved. Thus, synaptic plasticity can improve efficiency by allocating resources to deal with patterns that have been learned by experience.

Avoiding High Information Rates

A brain can also improve its energy efficiency by distributing its information among low-capacity neurons to avoid, whenever possible, cells transmitting at high rates. This strategy exploits the fact that information tends to be cheaper at lower rates, a fact that could help to explain why most neurons are small, low-rate cells. Ongoing studies suggest that the mammalian retina improves the energy efficiency for signal transmission along the optic nerve by sending 60% of its information down low-rate cells (Koch et al., 2006; Balasubramanian & Sterling, 2009).

Miniaturization

Shrinking circuits by making the components as small as possible is effective at reducing energy consumption. The trend in electronics is to improve energy efficiency by decreasing the size and increasing the density of components on chips. This reduces energy consumption by reducing the surface area, and, hence, the capacitance, of components, down to physical limits imposed by noise and heat production. Nervous systems, with their masses of tiny, densely packed neurons, have responded to similar pressures. Miniaturization reduces the number of ion channels needed to produce neural signals and the membrane capacitance that has to be charged by reducing membrane area. Synapses are typically about 1 μm across, and axons and dendrites have narrow calibers, typically 0.3 to 0.8 μm diameter in cortex (Braitenberg & Schütz, 1998). It is this high degree of miniaturization that enables the cortex to pack an average of 10^5 neurons, 10^8 synapses, 4 cm of dendrite, and 4 km of axon branches into 1 mm^3 of gray matter. Whether miniaturization has reached the level where it is limited by heat production has not been established, but it has approached the limits imposed by noise by pushing the diameters of axons to the limit imposed by thermal fluctuations in ion channels (Faisal, White, & Laughlin, 2005).

Wiring Efficiency

The most compelling evidence that nervous systems have evolved to be efficient comes from studies of wiring efficiency (Chklovskii & Koulakov, 2004; Laughlin & Sejnowski, 2003). Nervous systems reduce the lengths of their neuronal connections, and, hence, energy, material, and time costs, in a host of ways. The great Spanish neuroanatomist Santiago Ramon y Cajal observed at the end of the nineteenth century that the shapes of neurons were designed to reduce the length of their connections to conserve space, materials, and time. The modern era of research into neural wiring efficiency was initiated by a philosopher, Christopher Cherniak, and a mathematician, Graeme Mitchison (Cherniak, 1995; Mitchison, 1992). Cherniak was interested in bounded rationality and set out to test the proposition that, if a brain is resource limited, it will have evolved to use these resources efficiently. Cherniak took wiring minimization algorithms, sophisticated computer programs that engineers had developed to work out the optimum placement of components on computer chips, and applied them to one of the simplest and most thoroughly mapped nervous systems, the 302 neurons of the nematode worm *Caenorhabditis elegans*. Of the 40 million ways one could position

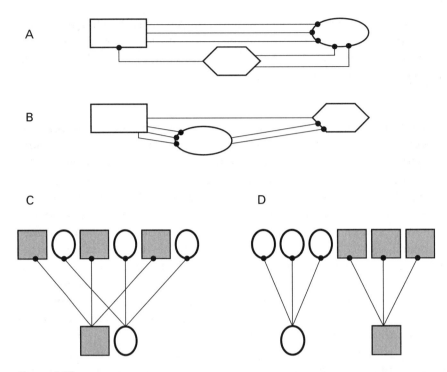

Figure 2.10

Reducing wiring costs by optimizing component placement (Cherniak, 1995) and by dividing up a large area into two specialized regions (Mitchison, 1992). In (A) and (B), four components (they could be ganglia or neurons) make different numbers of connections with each other. The large numbers of long connections in (A) are wasteful, and the total length of wiring is reduced by placing the more strongly interconnected components closer together (B). Dividing up one area in which two types of circuit are operating (C) into two separate areas, one for each circuit (D), reduces the total length of wiring because the neurons in one circuit do not have to make room for the neurons in the other circuit.

the clusters of neurons, the worm uses the most efficient. The fact that components can be placed in nervous systems to reduce the total length of wiring is easily demonstrated (figure 2.10A,B).

Mitchison considered the principles that govern the mapping of information within and among cortical areas. He deduced, among other things, that a brain reduces wiring costs by mapping information into discrete areas (figure 2.10C,D). Consider a visual area that contains neurons processing information on color and neurons extracting information on patterns of motion. The color neurons will connect exclusively with other color neurons to form color-processing circuits, and the motion neurons will form exclusive motion circuits. Intermingling these two types of circuits increases wiring lengths because the connections forming the color circuit have to work around motion neurons, and vice versa. Consequently, separating the neurons into two separate areas, one for color processing and another for motion, will reduce wiring costs. The savings achieved by localizing specific tasks to specific areas increases with the size of these areas. This may explain why, among mammals, the larger brains have more functionally distinct cortical areas. For example, the smallest mammals have between two and five cortical areas devoted to vision, covering, in total, less than 1 mm^2 of the cortical surface. By comparison, large primates have more than 30 such areas, covering more than 500 mm^2 of the cortical surface (Kaas, 2000). Recent studies have driven home the point that nervous systems are organized to minimize wiring costs. Efficient design is observed at all levels of anatomical organization, from the positioning of functional areas across the cerebral cortex to the arrangement of synaptic spines along dendrites (Chklovskii, 2004; Chklovskii & Koulakov, 2004). There is no doubt that the need for efficiency has left its stamp on the human brain.

Conclusions

We have demonstrated that brains must consume energy to function and that these requirements are sufficiently demanding to limit performance and determine design. The rationale is simple. We start with function, the reason why nervous systems process information. A nervous system directs work that is intended to increase an animal's fitness, where *fitness* is defined as lifetime reproductive success. To direct work effectively, a nervous system uses sensory receptors to collect information that specifies its situation—the state that the animal is in. A brain's neurons then process this information to generate the instructions to effectors in order that they do work that improves the animal's state. These instructions are also information, because, like the sensory receptors, the instructions determine a state; for example, the set of muscular contractions that coordinates the movement of a limb.

Brains collect, process, and distribute information to direct work. We have shown how it is possible to put this statement on a proper scientific footing by measuring the brain's commodity—information—in bits. Bits measure the ability to determine state in units of binary choice. Bits in identified neurons, namely fly photoreceptors, can be measured. It is important to emphasize that, although bits appear to be rather specialized entities that inhabit the entrails of computers, they can quantify all forms of information. Indeed, one can go further, like the physicist John Wheeler, who coined the phrase *It from Bit* to launch his proposition that information is universal because nothing exists in the universe without a determination of state. This fascinating suggestion is being vigorously pursued but, because the universe is large, complicated, and largely unknown, it has some ways to go. The same can be said of the application of information theory to the human brain.

A physical view of bits reveals another universal. Bits require energy, the energy required to set or sense a state. Thus, the transmission and processing of information depends on energy. Brains cannot violate this physical law—they, too, must apply some energy for every bit they use, but novel measurements on fly photoreceptors show that the energy a neuron uses is 10 million times the physical lower limit. This profligacy is not wanton sloppiness and waste; it stems from the need to instruct appropriate behavior on demand. The biophysics of neural signaling compels brains to pay a great deal extra to direct signals through circuits, and costs are greatly inflated by two desirable attributes of behavior: speed and accuracy.

Finally, the link between information and energy in the brain is much more than a neat bit of physics: It has teeth and it bites. Energy requirements limit processing power to the order of one signal per neuron per second in the human brain. Evolution's attempt to reduce the impact of this limit by increasing energy efficiency is readily apparent in the design of neurons and nervous systems. The most striking examples to date are evident in the fabric of brains, including the human brain. The dense packing of tiny cells with thin connections, the grouping of neurons into functional areas, and the orderly arrangement of areas save space, materials, and energy.

In conclusion, nervous systems use information to direct energy into work, and they use energy to direct information. But what good comes from the realization that the human brain, and brains in general, have to work for their livings? There are several answers. Energy efficiency will help us to piece together the brain's circuits and decipher its codes through a better appreciation of design factors and principles. For example, we have determined that energy efficiency demands sparse codes. Furthermore, a knowledge of the energy costs of neural function and their relationship to performance illuminates brain evolution (Niven & Laughlin, 2008). The reason is simple: Natural selection involves trade-offs between costs and benefits, with energy a major cost and information a benefit.

Returning to the theme of this book, we see that organisms use information to direct work in an attempt to ensure that energy is put to good use. But there is a trade-off. Although information processing saves energy by directing work more appropriately, information processing also uses energy to drive signals through circuits. These energy demands are particularly high in nervous systems, because these systems have to process large amounts of information quickly and reliably and transmit them over considerable distances. In this real sense, evolving animals resemble businesses, economies, governments, and societies. These organizations all attempt to strike a balance between two alternatives. Does one devote resources to process information for better management? Or is it better to use these resources to do more of the same—for example, increase the volume of production under existing management schemes? The fact that brains both expand and contract during evolution reflects the fact that the optimum balance changes with circumstances.

References

Aiello, L. C., Bates, N., & Joffe, T. H. (2001). In defense of the expensive tissue hypothesis. In D. Falk & K. R. Gibson (Eds.), *Evolutionary anatomy of the primate cerebral cortex* (pp. 57–78). Cambridge: Cambridge University Press.

Allman, J. M. (1999). *Evolving brains*. New York: Freeman.

Ames, A. (2000). CNS energy metabolism as related to function. *Brain Research Reviews, 34*, 42–68.

Attwell, D., & Laughlin, S. B. (2001). An energy budget for signaling in the grey matter of the brain. *Journal of Cerebral Blood Flow and Metabolism, 21*, 1133–1145.

Balasubramanian, V., & Sterling, P. (2009). Receptive fields and functional architecture in the retina. *Journal of Physiology-London, 587*(12), 2753–2767.

Bekenstein, J. D. (2003). Information in the holographic universe. *Scientific American, 289*(2), 58–65.

Braitenberg, V., & Schütz, A. (1998). *Cortex: Statistics and geometry of neuronal connectivity* (2nd ed.). Berlin: Springer.

Carroll, S. B. (2005). *Endless forms most beautiful: The new science of evo devo and the making of the Animal Kingdom*. New York: Norton.

Cherniak, C. (1995). Neural component placement. *Trends in Neurosciences, 18*(12), 522–527.

Chklovskii, D. B. (2004). Synaptic connectivity and neuronal morphology: Two sides of the same coin. *Neuron, 43*, 609–617.

Chklovskii, D. B., & Koulakov, A. A. (2004). Maps in the brain: What can we learn from them? *Annual Review of Neuroscience, 27*, 369–392.

Clarke, D. D., & Sokoloff, L. (1999). Circulation and energy metabolism of the brain. In G. J. Siegel, B. W. Agranoff, R. W. Albers, S. K. Fisher, & M. D. Uhler (Eds.), *Basic neurochemistry: Molecular, cellular and medical aspects* (6th ed., pp. 637–669). Philadelphia: Lippincott-Raven.

de Ruyter van Steveninck, R. R., & Laughlin, S. B. (1996). The rate of information transfer at graded-potential synapses. *Nature, 379*, 642–645.

Dusenbery, D. B. (1992). *Sensory ecology: How organisms acquire and respond to information* (1st ed.). New York: W. H. Freeman.

Faisal, A. A., White, J. A., & Laughlin, S. B. (2005). Ion-channel noise places limits on the miniaturization of the brain's wiring. *Current Biology, 15*(12), 1143–1149.

Hebb, D. O. (1949). *The organization of behavior*. New York: John Wiley & Sons.

Johnson, B. (2009, 4 May). Power failure: How huge appetite for electricity threatens internet's giant. *The Guardian*, p. 13.

Kaas, J. H. (2000). Why is brain size so important: Design problems and solutions as neocortex gets bigger or smaller. *Brain and Mind, 1*, 7–23.

Koch, K., McLean, J., Segev, R., Freed, M., Berry, M. J. I., Balasubramanian, V., & Sterling, P. (2006). How *much* the eye tells the brain. *Current Biology, 16*, 1428–1434.

Landauer, R. (1996). Minimal energy-requirements in communication. *Science, 272*(5270), 1914–1918.

Laughlin, S. B. (2001). Energy as a constraint on the coding and processing of sensory information. *Current Opinion in Neurobiology, 11*, 475–480.

Laughlin, S. B., de Ruyter van Steveninck, R. R., & Anderson, J. C. (1998). The metabolic cost of neural information. *Nature Neuroscience, 1*(1), 36–41.

Laughlin, S. B., & Sejnowski, T. J. (2003). Communication in neuronal networks. *Science, 301*, 1870–1874.

Leff, H. S., & Rex, A. F. (2003). *Maxwell's demon 2: Entropy, classical and quantum information, computing*. Bristol: Institute of Physics Publishing.

Lennie, P. (2003). The cost of cortical computation. *Current Biology, 13*, 493–497.

Levy, W. B., & Baxter, R. A. (1996). Energy-efficient neural codes. *Neural Computation, 8*(3), 531–543.

Lloyd, S. (2000). Ultimate physical limits to computation. *Nature, 406*(6799), 1047–1054.

Misner, C. W., Thorne, K. S., & Zurek, W. H. (2009). John Wheeler, relativity, and quantum information. *Physics Today, 62*(4), 40–46.

Mitchison, G. (1992). Axonal trees and cortical architecture. *Trends in Neurosciences, 15*(4), 122–126.

Nawroth, J. C., Greer, C. A., Chen, W. R., Laughlin, S. B., & Shepherd, G. M. (2007). An energy budget for the olfactory glomerulus. *Journal of Neuroscience, 27*(36), 9790–9800.

Niven, J. E., Anderson, J. C., & Laughlin, S. B. (2007). Fly photoreceptors demonstrate energy-information trade-offs in neural coding. *PLoS Biology, 5*(4), 828–840.

Niven, J. E., & Laughlin, S. B. (2008). Energy limitation as a selective pressure on the evolution of sensory systems. *Journal of Experimental Biology, 211*(11), 1792–1804.

Pierce, J. R. (1962). *Symbols, signals and noise*. London: Hutchinson.

Schrödinger, E. (1944). *What is life?* Cambridge, U.K.: Cambridge University Press.

Shannon, C. E. (1949). Communication in the presence of noise. *Proc. Inst. Radio Eng., 37*, 10–21.

Shannon, C. E., & Weaver, W. (1949). *The mathematical theory of communication*. Urbana, Illinois: The University of Illinois Press.

Shulman, R. G., & Rothman, D. L. (Eds.). (2004). *Brain energetics and neuronal activity: Application to fMRI and medicine*. Chichester: Wiley.

Sokoloff, L. (2004). Energy metabolism in neural tissues *in vivo* at rest and in functionally altered states. In R. G. Shulman & D. L. Rothman (Eds.), *Brain energetics and neuronal activity* (pp. 11–30). Chichester: Wiley.

Starting where energy–information trade-offs left off, Michael Lightner takes us on an exploratory tutorial through trade-offs that exist in the work of any system, whether living or manufactured. Using techniques in optimization he and others originally developed for the manufacture of computer chips, Lightner shows how all work processes, living or nonliving, are subject to constraints. He describes how different objectives trade off against each other (leading to trade-offs such as Laughlin's energy–information trade-offs, described in chapter 2) and how an extreme case of trading off yield against everything else produces a much-desired result in chip manufacture, and perhaps in living systems, called design centering. *He examines the inherent nature of* performance–yield trade-offs, *which affect work in biological systems to an extent heretofore unrecognized. Throughout the chapter, he shows how these concepts help to explain a unique and real-life combination of human-designed and living systems, operating under intense optimality constraints—the Pony Express.*

3 Performance–Yield Trade-offs in Work in Manufactured and Living Systems: Design Centering Looks at the World of Work

Michael Lightner

Introduction

Ever since organisms began working at the dawn of life (chapter 1), they have found themselves in a series of constant trade-offs between one aspect of their existence or another: food and safety, growth and safety, speed and accuracy, size and mobility. These trade-offs have become evident to generations of biologists, some of whom have asked the question, how well does nature reach a balance between these competing demands? Manufactured products and systems are likewise characterized by trade-offs: cost and speed, durability and weight, complexity and yield, yield and performance.

In living systems, the concept of *yield* is widespread: How many chicks hatched? How often does a hunt result in a kill? These are yields. If all chicks hatch, that implies 100% yield. If every fourth hunt results in a kill, that implies 25% yield. The same concept is prevalent in manufactured systems. How many cars are "lemons" out of every 100 manufactured? How many packages of a food product are underweight out of every 100 produced? These are also measures of yield.

A logical next question is, why might there be variation in the results of a hunt, the hatching of eggs, the manufacture of automobiles, or the automatic filling of a food package? One answer is that there is inherent variability, or lack of complete control, in both living and manufactured systems. In the case of hunting, weather, terrain, and the age of the hunters and the hunted all contribute to the variability in the success or failure of the hunt. With cars, it is not possible to control the manufacture of each individual component to 100% tolerance. In car manufacturing and every other system, there are many more factors that come into play. Both living and manufactured systems are subject to inherent, uncontrollable variability.

The idea of variability leads, in turn, to the question of the range over which something can vary. Indeed, there are constraints on almost every characteristic of every manufactured and living system. We typically capture these constraints with a range. For example, the constraints imposed on a manufacturing system might be that the product weight must be within 5% of the advertised weight or the fuel efficiency of a specific model of automobile must be within 10% of the advertised value. We can likewise look at characteristics of living systems and note, for example, that the weight, height, and running speed of a specific predator are within certain bounds.

Finally, in living and manufactured systems, we are often faced with the concept of performance. Miles-per-gallon could be a performance measure of an automobile, and miles-per-hour might be a performance measure of a cheetah. Increasing performance is often a goal of human designers for manufactured systems (chapter 6). Likewise, improving performance is often viewed as a desirable outcome in living systems.

This chapter explores trade-offs, yield and variability, constraints, and performance improvement (often discussed in terms of optimization or optimality) in some detail, introducing concepts and using examples from both manufactured and living systems. Our goals are to understand the similarities that may exist between living and manufactured systems with respect to such concepts as trade-offs and performance and to expand the tools with which we can model both living and manufactured systems.

Part I first explores the central concept of *constraints* on performance and the equally important but often overlooked concept of *yield*. These concepts are combined in the development of a technique used to increase the yield of manufactured systems, called *design centering*, which we then use as a method to understand yield and performance relationships in living systems.

Part II next explores the concepts of *single-* and *multiple-objective optimization*, as they have been developed for manufactured systems, and examines how the concepts apply to living systems. We introduce a realistic example system with severe constraints on performance, yield, and optimality that involves both human designs and living systems: Our example is the U.S. Pony Express and other "equine express" systems—and we use earlier research on these systems to explore the usefulness of optimization approaches to real-life work.

Part III then brings together the concepts from parts I and II to develop the fundamental importance of *performance–yield trade-offs* for understanding work in living systems. These trade-offs, understood and utilized in manufactured systems, form an essential foundation for understanding performance, yield, optimality, and variability in the work of living systems. Finally, we analyze the work of the Pony Express, and the earlier research on this subject, using our integrative understanding of performance–yield trade-offs.

Part I: Constraints, Parameter Variation, and Yield in Manufactured and Living Systems and the Power of Design Centering

There are often multiple ways to model living and manufactured systems. In engineering, mathematical models of systems are common. However, in order to focus on concepts in this chapter, we will limit the use of mathematical equations and represent our examples graphically wherever possible. These *graphical models* can be used to predict system characteristics or performance and, with those predictions, to try to improve or optimize various elements of performance. We will use this approach throughout this chapter. (Note that our use of a "system model" is different from the use of a "model system" in the life sciences.) In manufactured systems, developing an effective model is key to the design process in engineering (chapter 6).

Constraints

Constraints on systems can occur in many forms. Consider the example of a constraint on the weight of a manufactured system. Think of launching a product into space—for example, a scientific instrument on board a satellite. To accomplish this task, the instrument must weigh less than a certain amount so that we can afford to launch the product. As another constraint, the power requirement of the instrument must be less than a certain amount to be powered effectively throughout the life of the satellite. However, in both of these examples, in order to accomplish the goals of the space mission, the instrument must have a certain weight and consume a certain amount of power. Thus, we have both minimum and maximum constraints, or bounds, on weight and on power.

As we state the conditions for a scientific instrument in space, we are stating our *constraints* on the design of the instrument. In our example, so far, for an instrument to be acceptable to launch into space, it must meet weight and power requirements. Perhaps, in addition, the instrument must meet volume restrictions and be capable of taking a required measurement, such as the spectrum of a star—not an easy task. There are many constraints associated with this design. Meeting these constraints is key to the successful design of the instrument.

One effective way to capture constraints is by using a common mathematical description, known as inequalities. For example, we can say that the scientific instrument must weigh less than 10 kilograms (kg), must have a volume less than 1 cubic meter (m^3), and must consume less than 24 watts (W) of power. We can then express the constraints in a set of inequalities expressed as equations:

$$weight \leq 10 \text{ kg}$$

$$volume \leq 1 \text{ m}^3$$

$$power \leq 24 \text{ W.}$$

Furthermore, because the instrument must have some weight, volume, and power in order to function, we could place a lower bound on the parameters, which would, in turn, give a range:

$$0 \leq weight \leq 10 \text{ kg}$$

$$0 \leq volume \leq 1 \text{ m}^3 \qquad (3.1)$$

$$0 \leq power \leq 24 \text{ W.}$$

As a very different kind of example, one combining human-designed systems and living systems, consider the Pony Express in the United States, one instance of an "equine express" system (Minetti, 2003). An equine express uses horses and human riders to transmit time-sensitive information over great distances. When such a system is designed, the developers must consider many different kinds of constraints: the endurance of the rider, the physiological limits of the horse, the weather, the terrain, the possibility of hostile forces, the nature of the message, and the importance of it being delivered reliably, as well as many other factors, such as the ability to staff way stations and the food, forage, and supplies required to maintain them. As with the earlier example of building the scientific instrument to launch into space, describing the constraints on the equine express system will require the use of inequalities. Equine express systems are interesting because of the complexity of the constraints and because of the embedded combination of human design and biological work. Indeed, one could say that the equine express could be used as a model system for system modeling of work in living systems.

Design or Model Parameters

In developing any complex system, such as a scientific instrument, there are at least two levels of design choices. One involves high-level choices of general characteristics for the system, such as the specific measurement techniques to be used, the materials for the housing, or the communication protocols. These high-level choices can be considered to be system-level design decisions.

Once the system-level choices are made, the detailed design must then be undertaken, which involves a second level of design choices. The detailed design will involve choices of values for a number of parameters. These parameters are known as the *design parameters* of the system. The values of the constraints (discussed above) will be determined by these design parameters.

How does the addition of design parameters, or model parameters, change the description of the constraints? The change with respect to the form of equations is straightforward. The set of parameters is denoted by x. (In real-world design models, there is usually more than one parameter.) The constraints are then stated as a function of the set x as follows:

$$0 \le \text{weight}(x) \le 10 \text{ kg}$$

$$0 \le \text{volume}(x) \le 1 \text{ m}^3 \tag{3.2}$$

$$0 \le \text{power}(x) \le 24 \text{ W}.$$

Feasible Region

From our discussion of constraints, it is evident that there will be a range of design parameter values over which the specified constraints are satisfied. Outside of that range, the constraints are violated. This range of the parameters, x, over which the constraints are satisfied is called the *feasible region*.

For our purposes, it is useful to show a plot of the feasible region. We will consider two design parameters, $x1$ and $x2$. (In a real-world example, there are often many design parameters and, in turn, a complicated graph.) A feasible region for parameters $x1$ and $x2$ is shown in figure 3.1.

Parameter Variation or Uncertainty

In both living and manufactured systems, the design parameters cannot be precisely controlled or exactly measured. For example, in the equine express system, the time it takes a horse and rider to travel from station to station might vary widely, depending on conditions. Thus, there is inherent variation around any given set of parameters. In experimental measurements, it is common to note this by showing the error bars

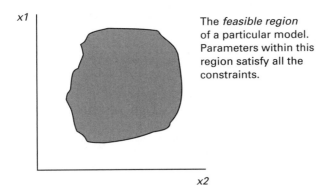

The *feasible region* of a particular model. Parameters within this region satisfy all the constraints.

Figure 3.1
The feasible region of a set of constraints.

associated with measurements. For instance, while we might say about a certain cell type that the thickness of the cell wall is 20 nanometers (nm), researchers in the field know that the cell wall thickness, even for the same cell type, will be 20 ± 0.2 nm.

Interestingly, this same type of variation is also common in manufactured systems. A bolt has a certain length and width and a variation of values about that length and width. At a far finer scale of manufacturing, an integrated circuit design, when turned into chips, will, even in the same fabrication plant, produce integrated circuits with a significant range of key parameters, such as speed.

For anyone working in a field that involves measurement, modeling with parameters, or manufacturing, the idea of *uncertainty* expressed above will be familiar. Consider what happens, though, when we combine the concept of constraints with that of uncertainty in parameters. Start with a set of parameters within the feasible region, a point in figure 3.1. Then plot the variation that the parameters may experience under uncertainty in the model under consideration. The result will be similar to that shown in figure 3.2.

Figure 3.2 shows that the region produced by parameter variation may have values both inside and outside the feasible region. The portion of the parameter variation area that overlaps with or falls inside of the feasible region, when divided by the total

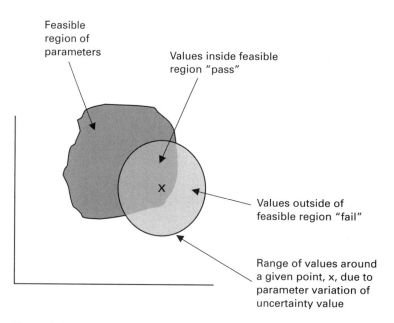

Figure 3.2
An illustration of parameter variation about a specific set of parameter values in the feasible region.

area of the parameter variation, provides the yield of the system. This is often expressed as a percentage yield. If only 15% of the area produced by parameter variation overlaps with the feasible region, then the yield of the system is 15%. If the variations are reduced in a way that increases the overlap to 85%, then the yield is 85%. (At a more technical level, the yield is defined in an engineering application by integrating the probability distribution function of the parameters, centered at point y, intersected with the feasible region over the entire probability distribution function.)

The concept of yield is a very powerful one for understanding the behavior of a system at work. Capturing constraints and the implied feasible region, understanding parameters and their variations, and measuring, calculating, or observing yield can lead to great understanding of both living and manufactured systems. However, in an engineering setting at least, understanding yield is not enough—one must understand how to *affect* yield.

Design Centering

In manufactured systems, from very large-scale integrated (VLSI) circuits to packages of cereal, low yield represents increased manufacturing cost. In living systems, low yield can represent decreased survival, or decreased reproduction, neither of which is desirable. Looking again at figure 3.2, it is possible to see that if we could move as much of the range of parameter variation as possible into the feasible region, we would increase the yield. This technique is called *design centering*, illustrated in figure 3.3.

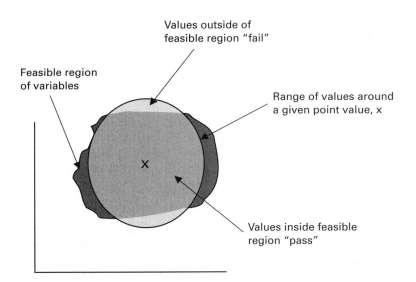

Figure 3.3
Moving the range of variations of x to maximize the overlap with the feasible region maximizes yield. This is known as design centering.

Design centering, which provides an intuitive, as well as computationally viable, method of yield maximization, has been the subject of much research (e.g., Abdel-Malek, Hassan, Soliman, & Dakroury, 2006; Ibbitson, Crompton, & Boardman, 1984; Lightner & Director, 1981b; Low & Director, 1991; Meehan, 1991; Pratap et al., 2006; Sapatnekar, Vaidya, & Sung-Mo, 1994; Zurada, Lozowski, & Malinowski, 1997). But what does it mean to "move" the range of parameter variations? In manufactured systems, it could mean using a different type of material or changing a component to a nominal value. In a living system, it could mean changing habitat (migrating) or building a different type of nest (chapter 5).

The Power of Design Centering

It is important at this point to pause and consider the power of these ideas. First, using models with parameters to describe both living and manufactured systems is common throughout science and engineering. Second, the requirement that certain constraints be met for a system to be viable or acceptable is also common. Examples include the temperature range within which an organism can survive or a device can operate, the range of time for viable gestation of mammals, and the maximum pressure survivable by certain organisms. More complex examples include the design and manufacture of a VLSI chip within fractions of a micron or the operation of the Pony Express across the vast span of the western United States. Third, it is not uncommon to have a range of performance that can be traced to a variation of model parameters. The range of model parameters does not necessarily provide a mechanistic rationale for variation, merely a direct link within the specific model.

We have thus captured a number of key elements of models with some frequently used concepts of constraints, parameters, and parameter variation. What is not as common, however, and what is vital to applying these concepts to living systems is to relate the concept of parameter variation to the yield of the system—that is, to the system's success or failure.

Moreover, once we understand the link between parameter variation and yield, we are led to possibilities for increasing the yield of the system. The graphical version of this concept is design centering, which presents a compelling conceptual model for designing systems with increased probability of success, as well as for exploring the behavior of systems, manufactured or living, and even for considering how the characteristics of systems may result from design-centering-like processes.

It would be possible to spend the rest of the chapter exploring consequences and possibilities around yield and design centering. Design centering has been explored in a number of manufactured systems, including VLSI chip design. However, it has not been explored in depth in living systems—we hope that readers will be motivated to take on such explorations. However, if we stopped at this point, with design centering, we would miss some equally key ideas for understanding ways that work gets done in living systems. The first of these is the general area of optimization of systems, the second is the area of optimizing systems with multiple competing objectives, and the third is optimization when one of the objectives is the yield of the system.

Part II: Optimization and Trade-offs in Manufactured and Living Systems

As discussed in part I, models that describe performance with constraints, parameters, and parameter variability are common. We showed how these led to the concept of the yield of a system, the desire to improve the yield of a system, and the desire to maximize the yield of a system. This step of maximizing yield, which has been illustrated with the approach of design centering, leads us, in turn, to the idea of optimizing system performance.

To examine the relationship between constraints and optimization, we will use the example of designing our satellite instrument. It is easier to take accurate measurements when the instrument uses more power. More power, though, increases the weight and the volume required for the instrument. But easier measurements might use more proven technology and, thus, reduce the cost and increase the accuracy of the instrument. However, larger and heavier instruments are much more costly to put into space. This set of relationships shows us that not only do we have *constraints* as part of the design, but we also have *trade-offs* operating among the various elements of the design. These trade-offs are *inherent,* in that they do not go away as long as the constraints are present. This concept of *inherent trade-off* will be key in the rest of this chapter.

The example of an instrument launched into space illustrates the relationship of constraints, optimization, and trade-offs introduced for a manufactured system. The same relationships are at work in a system involving an interesting combination of living creatures, human design, and both natural and manufactured constraints: the Pony Express in the United States. The Pony Express is one instance of an equine express system. When such a system is deployed, the designers must consider the endurance of the rider, the physiological limits of the horse, the weather, the terrain, the possibility of hostile forces, the nature of the message, and the importance of it being delivered reliably and in a timely manner.

For an equine express system, there is a clear goal: Deliver the message. However, it is almost always the case that there are specific temporal constraints: The message must be delivered within a short and specified period of time or the message is of no value. There are other clear constraints of a very different nature: the physiological limits of the horse and rider (Minetti, 2003). Similar to the way in which fundamental parameters underlie the constraints in our example of the satellite and its instrument, the limits on the horse depend, in turn, on the terrain, the weather, and the weights of the rider and message.

In addition, the design of an equine express system is complex and must adapt to the specifics of the local environment. For example, the more way stations that are built, the more costly the system and delivery of messages, because each way station requires staff and supplies, which often must be provided at great expense or cost in natural resources in uncompromising terrain. Were we the designers of such a system, we could look at minimizing the cost of such a system under a set of constraints. Furthermore, we could identify trade-offs among these constraints. But, just as the scientific instrument had a single raison d'être—to take a measurement—the equine express had the goal of always delivering the message, within a specified time. If the message was not delivered, the system failed. If the measurement is not made, the instrument fails. We know from part I that we can capture the successes and failures as the yield of the system. Thus, even when we are optimizing for a specific goal, yield always plays a critical role.

This discussion illustrates that we must consider three scenarios as we proceed:

1. optimization of systems with a single objective and a set of constraints;
2. optimization of competing objectives, called *multiple objective optimization*; and
3. trade-offs between performance and yield in part III.

Single-Objective Optimization

Mathematical treatments of optimization abound; here, we use a graphical treatment to present the concept. Chong & Zak (2008) provide a more detailed introduction to optimization.

The first step in our graphical approach will provide a straightforward starting point from which we can build. Suppose you had to find the minimum of some characteristic of a system. Furthermore, suppose that the characteristic of the system is represented by the curve in figure 3.4. The question is, what value on the horizontal axis (x-axis) will give the minimum value of the vertical axis (y-axis)? Examining figure 3.4 shows that the minimum, which is zero, occurs when x is equal to 2. If our objective is to minimize the function of x represented by this curve (which might, for example, be the minimum energy configuration of a system as a function of location), then we have successfully solved an optimization problem, and we have done so using a graphical approach.

Although one can imagine that most optimization problems have more variables and more complexity than this example, the image of figure 3.4 accurately represents the formulation of an optimization problem. More complex problems, of course, require far more complex pictures, but optimization problems can usefully be represented in this way.

A technical note before proceeding: Throughout the remainder of the chapter, we will be considering various examples with goals of *optimizing* some process. In math-

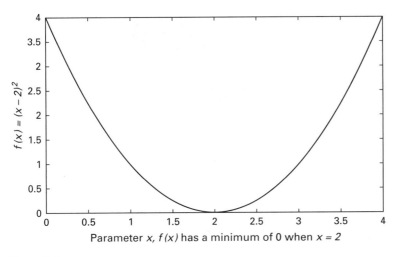

Figure 3.4
Example plot of a single-objective function with no constraints.

ematical terms, optimizing something is always expressed as either *minimizing* or *maximizing* a given objective. The objective to be optimized is sometimes referred to as an *objective function*. The use of the terms *optimizing, minimizing,* and *maximizing* can at times be confusing. The following fact often eliminates much confusion: Minimizing and maximizing are both examples of optimizing. Moreover, it is easy to turn a minimizing problem into a maximizing one and vice versa. Although some optimization papers, texts, and examples concentrate on "min," and others on "max," it makes no fundamental difference. Therefore, in the remainder of the chapter, we will move between these two types of optimization problems as best suits a given example.

We spent much of the first part of the chapter describing constraints and feasible regions. How do these come into play in optimization?

Let us begin with a single constraint. We will assume that there is one parameter, x, in the description of the system. The constraint is that

$$x \geq 3. \tag{3.3}$$

The constraint, in turn, defines a feasible region. Having identified this constraint, our first step in finding the optimal constrained solution will be to plot the feasible region shaped by this constraint, which we do in figure 3.5. (The shaded feasible region would continue indefinitely to the right of the figure, for all values of x greater than or equal to 3 are feasible.)

As the second step, we will combine the plot of the objective function shown in figure 3.4 with the feasible region shown in figure 3.5 to get the intersection of the

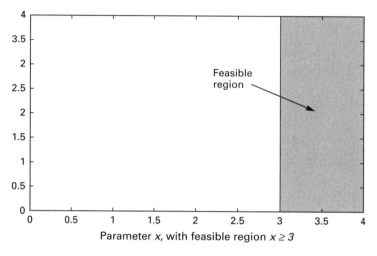

Figure 3.5
Plot of the feasible region given by equation (3.3).

objective function with the feasible region, shown in figure 3.6. The constrained minimization problem asked for the smallest value of the objective function from values of x within the feasible region. Thus, we see that the minimum value, which is 1, occurs when x is equal to 3.

We have already introduced systems with multiple constraints in part I. As we start to consider multiple parameters and multiple constraints, the capability to represent problems with graphical representation becomes limited, and we must use mathematical descriptions. However, the concepts are still the same:

1. Find the region of the parameters represented by each constraint.
2. Take the intersection of these regions to find the feasible region.
3. Find the minimum value of the objective using parameter values within the feasible region.

These steps appear easy but are complicated to achieve. Straightforward examples will illustrate this point.

We now come to the question of trade-offs, which are also referred to as competing objectives—competing multiple objectives.

Multiple Objectives
The examples thus far have been straightforward—perhaps artificially so. Of course, you might say, we would want to minimize the delivery time of the Pony Express. But we would also want to minimize the cost, and we would definitely want to maxi-

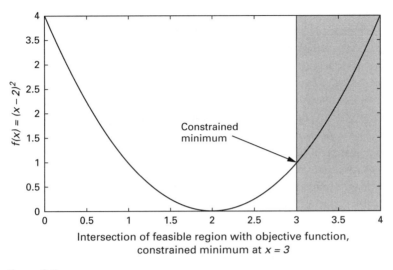

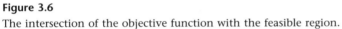

Intersection of feasible region with objective function,
constrained minimum at *x = 3*

Figure 3.6
The intersection of the objective function with the feasible region.

mize the probability that the message is delivered. Likewise, for the scientific instrument to be launched into space, certainly we would want to minimize weight. But, at the same time, we somehow also want to maximize the instrument's sensitivity, as well as the amount of data that can be stored and the lifetime of the instrument when it is in space.

In other words, the idea of a single objective for a design does not seem to match what is really happening in a manufactured or a living system. This has been recognized in many fields. One way to make the statement of the optimization problem more realistic is to say that we have several objective functions, which we want to optimize simultaneously. This approach is used in many fields under several names: multiobjective optimization, multiple criterion optimization, multiple-objective optimization, even multiple criterion decision-making.

There is one key question asked at the outset of formulating a multiple-objective optimization problem: Do the various objectives compete? For example, if there are two objective functions to be optimized, and the minimum of one function is consistent with the minimum of the other function (or identical, if the minima are unique), then they do not compete. If this were the case, then we need only choose one of the objectives and optimize it; the other objective will come along for free.

For our purposes, the real interest comes when we realize that our objective functions are competing. In real life, this is to be expected, as shown in our previous examples. For example, increasing instrument sensitivity requires more power, which, in turn, requires more weight or more volume. However, we cannot minimize volume and maximize instrument sensitivity at the same time—we must trade off between them. Our experience in manufactured and living systems tells us that most will have competing objectives.

Consider the ever-present contemporary concern with increasing the fuel efficiency of an automobile. One way to increase fuel efficiency is to reduce the weight of the car. However, we are also concerned with safety. Improving safety often involves such modifications as adding internal supports to the doors of the car or adding heavier bumpers—in other words, increasing the weight. At some point in designing a car, increasing the gas mileage (reducing the weight) will cause the safety to decrease. We then have a trade-off between mileage and safety.

The trade-off between mileage and safety can be considered in a graph; see figure 3.7. (Note that figure 3.7 is drawn simply for illustration not to represent the actual behavior of a car.) We will consider that parameter w is the weight of the car. Increasing weight decreases mileage and increases safety. figure 3.7 illustrates the trade-off.

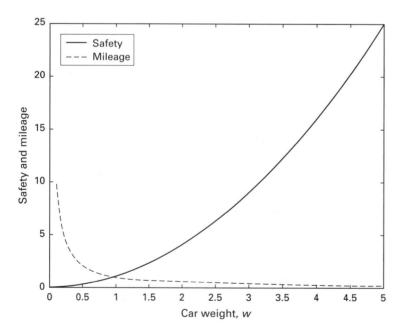

Figure 3.7
Trade-off illustrated by a plot of automobile mileage and safety, with car weight as the parameter.

In figure 3.7, the horizontal axis represents weight. The vertical axis has both a measure of safety and a measure of fuel efficiency plotted. The dashed line represents a measure of fuel efficiency—we see that, as the weight increases, the fuel efficiency decreases. The solid line represents a measure of safety—we see that, as the weight increases, the safety increases. Notice that there is a point where the graphs cross. If the two objectives are equally important, the best combined performance is shown at the crossing point—an equal trade-off. However, suppose that fuel efficiency was twice as important as safety for a given design. Then we would find our best trade-off at a different weight—a different point on the curves.

This discussion raises a critical point: What is the relative importance of the various competing criteria? In different situations, that relative importance may change, and, thus, the best trade-off will change. This means we have a family or set of best optimal solutions! In other words, it is not meaningful to talk about a single "best" solution without taking into account the different situations at hand and how these affect the relative importance of competing criteria.

A starting statement of an optimal solution to a trade-off problem is that it is any set of parameters that satisfies the constraints—that is, which is feasible—but for which improving one objective leads to poorer performance in another objective.

In multiple-objective optimization, a set of such trade-off solutions is called *Pareto optimal points*. The term is named after the nineteenth century French–Italian sociologist, economist, and philosopher, Vilfredo Pareto, who developed the concept. Pareto optimal points represent a surface in the space of the competing functions that contains the set of best optimal solutions under different conditions. This surface is called the *Pareto surface* or the *efficient frontier*.

Giving a graphic representation of such trade-offs requires another step. In figure 3.7, we plotted each of the competing criteria against the parameter. Plotting the Pareto surface requires that we plot the two objectives (in figure 3.7, mileage and safety) against each other. To do so, as in figure 3.8, we take every set of parameters in the feasible region, evaluate the objectives at each point, and plot these on a graph where each of the two axes is the value of a different objective. This is much easier shown than it is to describe: figure 3.8 shows the possible values of mileage and safety plotted against one another. Every point on the curve in figure 3.8 is a feasible trade-off point, or a Pareto solution. The entire set of points, the curve, is the Pareto surface or the efficient frontier. It represents the entire feasible set of best optimal solutions to the trade-off between mileage and safety. Which one of these points might be best can only be identified for a given situation and the relative weights, related to that situation, placed on mileage and safety. In other words, the best solution for mileage is not the best solution for safety. This trade-off needs to either be shown explicitly—the Pareto surface—or extra information, usually in the form of specific

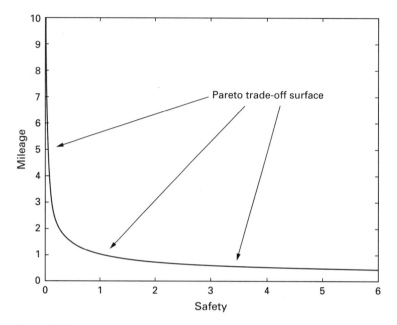

Figure 3.8
Plot of a portion of the Pareto surface, or the set of optimal trade-offs, of mileage (fuel efficiency) versus safety, developed from figure 3.7.

priorities or weights between the competing functions, needs to be introduced from the outside to yield a single solution.

This type of problem occurs and is addressed in economics and business on a regular basis. Often it is couched in terms of decision-making: making the best decision or choice in the face of competing objectives and spawning the related field of *multiple criterion decision-making* (e.g., Cortelazzo & Lightner, 1984, 1985; Jiguan, 1979; Lightner & Director, 1981a; Qian, Li, & Liu, 2002; Xiaohui & Eberhart, 002); there are several ways of solving for portions of the trade-off surface.

Multiple-Objective Optimization under Constraints
Just as realistic single-objective optimization problems have constraints on the parameters, so do multiple-objective optimization problems. Understanding the idea of constraints on the system from part I and the idea that those constraints can be viewed as a feasible region of the parameters, as in figure 3.5 are two foundations for understanding the effect of constraints on multiple-objective optimization. Figure 3.8 shows only the optimal trade-off surface. More generally, though, the possible objective values will be a region in a graph, analogous to the feasible region,

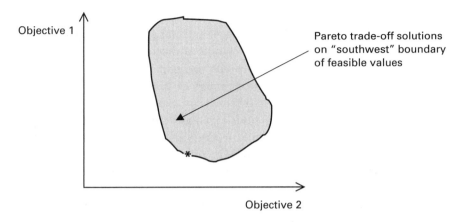

Figure 3.9
The shaded area represents the possible feasible values for the two objectives. If we are minimizing both objectives, then the lower left portion, or "southwest" portion, of the curve represents the efficient frontier of trade-off solutions.

and similar to figure 3.8. (A mathematical description of this relationship is available in Cheney & Light, 2009). To create a more realistic understanding of possible solutions, we will plot in figure 3.9 all the feasible values of two objectives, which, in turn, creates a region, not simply the optimal solutions that were represented in figure 3.8.

If we think about the problem of simultaneously minimizing Objective 1 and Objective 2 in figure 3.9, we notice two things. First, in the middle of the shaded region, we can simultaneously decrease both objectives—they are not competing there. This has implications in terms of design centering and real life. Second, in the lower left or "southwest" boundary of the shaded region, at the point marked with an asterisk (*), to then decrease Objective 2 and to stay in the shaded region, we must *increase* the value of Objective 1. In other words, the asterisk is a Pareto optimal point. Moreover, any point on the southwest boundary is also a Pareto optimal point; thus, we can see the entire trade-off surface of feasible values.

Alternative Approaches to Modeling Using Multiple-Objective Optimization
There are various systems whose behaviors have been modeled with multiple objectives. The category that we will review began with an approach called *particle swarm optimization*. The work of Kennedy and Eberhart (1995) attempted to use patterns in the natural world, such as flocking, schooling, and swarming together with ideas from evolutionary computation, for the purpose of modeling human social behavior. In this setting, we clearly have competing objectives. At the very least, the objectives of

individuals' goals and behaviors are quite likely to conflict with the objectives of group goals and behaviors. Kennedy and Eberhart developed simple behavioral rules and were able to both model and optimize various commonly occurring problems in social systems. These systems do not construct a model of global behavior; rather, they formulate simple local rules about how the elements or particles of the system will change from one moment to the next, when there is an overall objective common to all the particles. For instance, a flock of birds can form from each individual bird having an objective of reducing the amount of work done by the individual by reducing drag while flying. Running the optimization model results in identifying the "behavior" of a large number of particles, each following a given objective. Flocking, in this model, is therefore a result of optimizing an individual objective over a population. These behaviors, identified by running a particle swarm optimization model, satisfy the goal of optimizing an objective or objectives and, interestingly, mimic the behavior of numerous living systems.

Coello, Pulido, and Lechuga (2004) extended the work of Kennedy and Eberhardt by showing how multiple objectives could also be modeled with particle swarm optimization. They were working in an area now called *evolutionary multiobjective optimization*. They use the idea of particle swarm optimization as a technique for exploring an entire feasible region to find the Pareto solutions within.

In both of these examples, we are working outside the framework that we have set up in this chapter. Some of the ideas from these approaches are similar, but the models are often different. There are many excellent references in these areas for those interested in exploring these approaches further (e.g., Eberhart & Shi, 2004; Kennedy & Eberhart, 1995, 1997; Ozcan & Mohan, 1999; Poudyal, 1997; Pratap et al., 2006; Renato & Leandro dos Santos, 2006; Rubio del Saz, Gutierrez Blanco, Saez de Adana, & Catedra Perez, 2007; Shi & Eberhart, 1998, 1999; Shi, Eberhart, & Chen, 1999).

The concepts of competing objectives are key in modeling almost any system. An important point is that when considering optimization of either manufactured or living systems, it is critical to look at competing objectives and map out the sets of optimal trade-off solutions. Lacking other data to provide specific preferences or weights, all optimal trade-off solutions are equally valid solutions of the multiobjective optimization problem. It is very important that we maintain a multiobjective framework to fully understand the possible solutions available through setting specific priorities or weights.

Part III: Performance–Yield Trade-offs in Manufactured and Living Systems

We have thus far described systems made up of sets of constraints, which are, in turn, controlled by parameters. The parameters themselves are subject to variation or uncertainty. Exploring the uncertainty of parameters led us to the concept of yield, which led, in turn, to the concept of design centering as a mechanism to increase yield. Next, we added a performance objective that was to be optimized within the limits of the constraints. Finally, we explored the very common real-life situation of having two or more competing objectives. The search for solutions to competing objectives introduced the concept of optimal trade-off solutions and the Pareto surface or efficient frontier.

We have not, however, yet asked: Are there any special challenges if one of the competing objectives in a multiobjective optimization is itself the yield of the system? This is the last part of our puzzle. The answer to this question ties together the preceding content in this chapter and allows us to apply techniques developed for manufactured systems, such as design centering, to the work of living systems.

Uncertainty, Yield, and Optimization

In our discussion of design centering, we introduced the concept of moving the design point toward the center of the feasible region. Centering the design point will increase the yield of the system. It does so by decreasing the impact that parameter variation has on the yield of the system (figure 3.3). Centering the design point, therefore, is the equivalent of optimizing the system around a single objective: yield.

What if there were additional objectives other than yield? For example, in an equine express system, design centering—maximizing the yield—would maximize the probability that a message would reach its destination successfully and be delivered. However, by maximizing yield, the competing goal of minimizing the time to deliver a message is ignored completely. Thus, we are faced with a trade-off between yield and other objectives. We need to find a solution if this technique is to be useful in the real world, whether guiding manufacturing processes or explaining living systems.

We will explore this question by asking about the impact of parameter variation on optimal trade-off solutions. We know that parameter variability leads to a range of possible parameter values and, thus, a range of possible objective values. In figure 3.9, we developed the framework of a feasible region of solutions, bounded by the efficient frontier of Pareto trade-off solutions. We will use that framework as the basis for constructing figure 3.10, which adds the effect of parameter variations from figure 3.9, to demonstrate the effect on the efficient frontier.

Now the trade-off of performance and yield can become clear: Optimal trade-off solutions must always be at a boundary of a region, as in figures 3.9 and 3.10. If our parameters have uncertainty, however, which they must, then some of the resulting solutions that are generated will not satisfy the constraints on the system. They will fail and, therefore, lead to reduced yield.

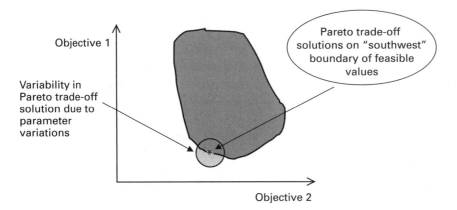

Figure 3.10
Variability about an optimal trade-off point. Values of the objectives that fall outside of the darker shaded regions due to parameter variability fail and thereby lower the yield of the system.

We might say that performance–yield trade-offs are *inherent* in any system produced by a process akin to manufacturing. That could be a computer chip; a living system developed, for instance, in an egg; or an equine express system designed by humans and shaped by the biotic requirements of horses and riders. In a living system, one direct effect of such trade-offs would be that a system may not make important transitions in life history. For example, a chick embryo may not successfully make the transition from egg to chick (Bekoff, 1992) and will not survive, or a blue tit may not make the transition from flightless to flight, and thus be highly unlikely to continue developing. The connection between the transitions in life history and optimization is based on the notion that the objective of any particular life phase is to successfully transition to the next phase; that is, to increase the yield of moving from one life phase to the next. The trade-offs of optimization of performance and yield are well documented (e.g., Brayton, Director, & Hachtel, 1980; Duenas & Mort, 2002; Hassan, Abdel-Malek, & Rabie, 2003; Hocevar, Lightner, & Trick, 1983, 1984; Lightner & Director, 1981b; Marseguerra, Zio, Podofillini, & Coit, 2005; Mezhoudi & Poudyal, 1997; Poudyal, 1997).

Now it is possible to illustrate the trade-off that occurs between performance and yield. First, consider that if we ignore yield as an objective, then the solution to a multiple-objective optimization problem will be the Pareto surface introduced in figure 3.9. For each optimal trade-off solution, the parameters will have some variation, as illustrated in figure 3.10. This will lead to a set of solutions about each individual trade-off solution. Many of the solutions, due to parameter variation, will fall outside the feasible region and thus represent a failure. This is illustrated in figure 3.11.

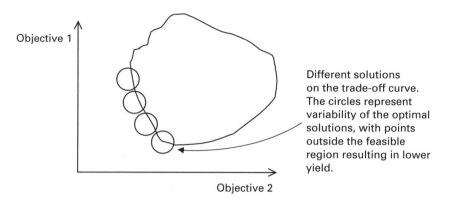

Different solutions on the trade-off curve. The circles represent variability of the optimal solutions, with points outside the feasible region resulting in lower yield.

Figure 3.11

Each circle represents the variability of a trade-off solution at its center due to parameter variation. Many of the potential solutions fail.

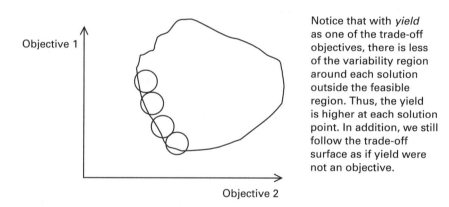

Notice that with *yield* as one of the trade-off objectives, there is less of the variability region around each solution outside the feasible region. Thus, the yield is higher at each solution point. In addition, we still follow the trade-off surface as if yield were not an objective.

Figure 3.12

As we consider yield as an objective, the circles, representing variability, tend to move inside the feasible region and away from the efficient frontier. Thus, yield increases, but performance decreases.

If, however, we consider yield to be important, as is always the case to a greater or lesser extent in both manufactured and living systems, then a trade-off between yield and performance becomes inherent. The solutions to this particular kind of multiple-objective problem, between competing objectives of performance and yield, will move away from the Pareto surface, the efficient frontier, so that more of the range of the solutions that are created by parameter variation will fall within the feasible region rather than outside the region. Making this change increases the yield, at the expense of performance, and is illustrated in figure 3.12. The extreme case of this trade-off would be to only consider yield, which produces design centering, in the same way as shown earlier in figure 3.3.

Full Circle: The Equine Express System

Now we will take a closer look at the equine express system in light of the ideas developed throughout this chapter related to trade-offs, performance, and yield. As stated earlier, the equine express is a horse-based communication system, primarily for transportation of low-weight, high-value items by a series of couriers. The U.S. Pony Express system is likely the best-known, but these systems have existed over the centuries in many locations, including Persia and China. Minetti (2003) studied a number of these systems and concluded that the spacing between rest stops was optimized at the limit of equine energy expenditure, both within a given equine express system and across such systems.

Taking that analysis as a starting point, let us continue to see what more information we can glean by analyzing this system using the techniques in this chapter. It is clear that the spacing of rest stops indeed cannot exceed the limit of equine energy expenditure. This is an inherent constraint on a physical parameter of the kind we described at the outset of the chapter. However, if the spacing actually were at the limit of equine energy expenditure, we would have a system with little or no *robustness*: Any effective equine express must be highly robust, which means it must be tolerant to variations in parameters. This means the system has high yield, because it is essential for the messages or packages to get through with high reliability.

In turn, the concept of a performance–yield trade-off indicates that to get this high yield, performance parameters will need to be "backed off": The performance parameters will need to be shifted in exactly the kind of shift used in design centering (figure 3.3). In order to increase the performance yield of the horses and their riders getting through to very close to 100%, the performance parameter of the length of ride for a horse must be moved away from the constraint boundary, which, in this case, is the physiological limit of the horse, toward a length that guarantees that close to 100% of the horses and riders will be able to ride in all conditions without exhaustion. Thus, we would expect to see a ride distance substantially less than the physiological limit of the horse.

Indeed, in a brief reexamination of Minetti's data and some discussion with long-distance horse riders, some histories of the Pony Express, and discussions with equine express buffs in various parts of rural Kansas, we find what our analysis throughout this chapter predicts. In fact, there is wide variability in the spacing of equine express way stations, and therefore of ride lengths, which vary by the kind of terrain encountered, the type of package carried, and the historical purpose of the express system. Undoubtedly there are other variables, just as in the examples earlier in the chapter.

However, the spacing of way stations and the length of the ride are not fixed by the capacity of the horse, but instead are design centered by a performance–yield trade-off. The equine express system, then, is an example of the real-world effect of design centering and performance–yield trade-offs on an optimization problem that not only combines multiple competing objectives, but combines manufactured, or human designed, systems with living systems.

Moreover, what drives this trade-off is the importance of *information yield* to this system. We can therefore propose an additional connection to themes in this volume. Equine express systems—in particular, high-speed systems like the Pony Express—are also highly energy intensive, in the amount of energy that must be used to fuel horses, riders, and way station staffs. The higher the amount of energy invested, the higher the reliability of information, for the greater the number of horses, riders, and staff, and the higher the yield in terms of information that successfully transits the system.

Thus, the particular performance–yield trade-off identified in the Pony Express system may be considered as a macroscopic example of the principle introduced by Laughlin of an inherent energy–information trade-off (chapter 2). Obtaining high amounts of information from the system requires investments of large amounts of energy. In a very practical way, shortening the length of a horse's ride to increase yield results in increased numbers of way stations. Every way station requires substantial energetic and monetary investment in construction, maintenance, staffing, and in fuel, including feed for horses and food for riders and station staff. In fact, very high-information-yield systems like the Pony Express are characterized by fantastic investments in energy and infrastructure, energetic requirements that ultimately leave them vulnerable to competition and shifting priorities. Indeed, the Pony Express, for all its fame, ran successfully for only 19 months.

What other issues need to be considered in such a system? Every horse and rider combination has a different energy balance. Weather plays a significant role, as does the terrain to be covered. Finally, the ability to recover from errors and emergencies, such as flash floods, wrong turns, landslides, and attack from hostile forces, is critical if there is to be a guarantee to delivering important messages, such as command decisions to armies.

In the end, the design of an equine express system requires robustness, which means it must be designed for high yield. Doing so requires careful consideration of several factors, not the maximization of one factor above all. These considerations all center, as illustrated above, on operating the system so that none of the factors are at their extremes. Our analysis thus differs from the conclusions of the original analysis, which focused on the optimization of a single objective, the physiology of the horse (Minetti, 2003). We argue that much could be gained by considering both living and manufactured systems from the perspective of inherent performance–yield trade-offs.

Summary and Suggestions

In a volume about work in living systems, this chapter has offered an alternative way to think about those systems, from the starting point of techniques developed for manufactured systems but applicable to work in living systems. Specifically, we considered the constraints that are placed on characteristics of a system. These constraints depend on parameters, which, in turn, are subject to variability or uncertainty. We thus arrived at the important concept of the yield of a system. Given the importance of yield, we were naturally tempted to explore ways of increasing or improving the yield and thus developed and reviewed the concept of design centering. Improving only yield was not enough, though; we also wanted to optimize certain measures of the system performance. We found that it was common in real-life manufactured or living systems to have several performance measures that were competing with one another, which led us to examine the ideas of multiple-objective optimization. Finally, we realized that in our Faustian desire to optimize performance, we were forgetting our initial concerns for the yield of the system, and examined the trade-off between performance and yield. At several points, our examination has led us away from a view that single-objective optimization of performance is likely to realistically characterize the operation of living systems any more than it does manufactured systems. Instead, the work of living systems can be profitably characterized and understood as systems that implement performance–yield trade-offs. Of such trade-offs, we believe the energy–information trade-off introduced by Laughlin in chapter 2 is a particularly important example of a performance–yield trade-off in the work of living systems.

The ideas of optimization, trade-off solutions, uncertainty, robustness, and yield are everywhere in the manufactured and living environments. These concepts provide a very useful framework for considering these systems and for broadening our discussion of systems, their design, and behavior.

References

Abdel-Malek, H. L., Hassan, A. S. O., Soliman, E. A., & Dakroury, S. A. (2006). The ellipsoidal technique for design centering of microwave circuits exploiting space-mapping interpolating surrogates. *IEEE Transactions on Microwave Theory and Techniques*, 54(10), 3731–3738.

Bekoff, A. (1992). Neuroethological approaches to the study of motor development in chicks: Achievements and challenges. *Journal of Neurobiology*, 23(10), 1486–1505.

Brayton, R., Director, S., & Hachtel, G. (1980). Yield maximization and worst-case design with arbitrary statistical distributions. *IEEE Transactions on Circuits and Systems*, 27(9), 756–764.

Cheney, E. W., & Light, W. (2009). *A course in approximation theory*. Providence, RI: American Mathematical Society.

Chong, E. K. P., & Zak, S. H. (2008). *An introduction to optimization* (3rd ed.). Hoboken, NJ: Wiley-Interscience.

Coello, C. A. C., Pulido, G. T., & Lechuga, M. S. (2004). Handling multiple objectives with particle swarm optimization. *IEEE Transactions on Evolutionary Computation, 8*(3), 256–279.

Cortelazzo, G., & Lightner, M. (1984). Simultaneous design in both magnitude and group-delay of IIR and FIR filters based on multiple criterion optimization. [see also IEEE Transactions on Signal Processing]. *IEEE Transactions on Acoustics, Speech, and Signal Processing, 32*(5), 949–967.

Cortelazzo, G., & Lightner, M. (1985). The use of multiple criterion optimization for frequency domain design of noncausal IIR filters. [see also IEEE Transactions on Signal Processing]. *IEEE Transactions on Acoustics, Speech, and Signal Processing, 33*(1), 126–135.

Duenas, A., & Mort, N. (2002). *Solving a multiple criteria decision-making problem under uncertainty.* Paper presented at the 2002 First International IEEE Symposium on Intelligent Systems.

Eberhart, R. C., & Shi, Y. (2004). Guest editorial special issue on particle swarm optimization. *IEEE Transactions on Evolutionary Computation, 8*(3), 201–203.

Hassan, A. S. O., Abdel-Malek, H. L., & Rabie, A. A. (2003). *Yield optimization via trust region and quadratic interpolation algorithm.* Paper presented at the 46th IEEE International Midwest Symposium on Circuits and Systems, 2003. MWSCAS '03.

Hocevar, D. E., Lightner, M. R., & Trick, T. N. (1983). A study of variance reduction techniques for estimating circuit yields. *IEEE Transactions on Computer-Aided Design of Integrated Circuits and Systems, 2*(3), 180–192.

Hocevar, D. E., Lightner, M. R., & Trick, T. N. (1984). An extrapolated yield approximation technique for use in yield maximization. *IEEE Transactions on Computer-Aided Design of Integrated Circuits and Systems, 3*(4), 279–287.

Ibbitson, I. R., Crompton, E., & Boardman, D. (1984). Improved statistical design centering for electrical networks. *Electronics Letters, 20*(19), 757–758.

Jiguan, L. (1979). Multiple-objective optimization by a multiplier method of proper equality constraints—Part I: Theory. *IEEE Transactions on Automatic Control, 24*(4), 567–573.

Kennedy, J., & Eberhart, R. (1995). *Particle swarm optimization.* Paper presented at the IEEE International Conference on Neural Networks, 1995. Proceedings.

Kennedy, J., & Eberhart, R. C. (1997). *A discrete binary version of the particle swarm algorithm.* Paper presented at the IEEE International Conference on Systems, Man, and Cybernetics, 1997. "Computational Cybernetics and Simulation."

Lightner, M., & Director, S. (1981a). Multiple criterion optimization for the design of electronic circuits. *IEEE Transactions on Circuits and Systems, 28*(3), 169–179.

Lightner, M., & Director, S. (1981b). Multiple criterion optimization with yield maximization. *IEEE Transactions on Circuits and Systems, 28*(8), 781–791.

Low, K. K., & Director, S. W. (1991). A new methodology for the design centering of IC fabrication processes. *IEEE Transactions on Computer-Aided Design of Integrated Circuits and Systems, 10*(7), 895–903.

Marseguerra, M., Zio, E., Podofillini, L., & Coit, D. W. (2005). Optimal design of reliable network systems in presence of uncertainty. *IEEE Transactions on Reliability, 54*(2), 243–253.

Meehan, M. (1991). *Understanding and maximising yield through design centering [microwave circuits].* Paper presented at the IEEE Colloquium on Computer Based Tools for Microwave Engineers.

Mezhoudi, M., & Poudyal, V. (1997). *A new method of yield maximization.* Paper presented at the 40th Midwest Symposium on Circuits and Systems, 1997. Proceedings.

Minetti, A. E., (2003). Efficiency of equine express postal systems: Relay riders over two millennia delivered mail with a remarkably consistent alacrity. *Nature, 426*(6968), 785–786.

Ozcan, E., & Mohan, C. K. (1999). *Particle swarm optimization: surfing the waves.* Paper presented at the Congress on Evolutionary Computation, 1999. CEC 99. Proceedings.

Poudyal, V. (1997). *On the center-of-gravity method of yield maximization.* Paper presented at the 40th Midwest Symposium on Circuits and Systems, 1997. Proceedings.

Pratap, R. J., Sen, P., Davis, C. E., Mukhophdhyay, R., May, G. S., & Laskar, J. (2006). Neurogenetic design centering. *IEEE Transactions on Semiconductor Manufacturing, 19*(2), 173–182.

Qian, F., Li, Q., & Liu, D. (2002). *Multiple-objective optimization decision-making method for large scale systems.* Paper presented at the 4th World Congress on Intelligent Control and Automation, 2002. Proceedings.

Renato, A. K., & Leandro dos Santos, C. (2006). Coevolutionary particle swarm optimization using Gaussian distribution for solving constrained optimization problems. *IEEE Transactions on Systems, Man, and Cybernetics. Part B, 36*(6), 1407–1416.

Rubio del Saz, A., Gutierrez Blanco, O., Saez de Adana, F., & Catedra Perez, M. F. (2007). *Swarm particle optimizacion apply to the searching of reflection points over NURBS surfaces.* Paper presented at the 2007 IEEE Antennas and Propagation International Symposium.

Sapatnekar, S. S., Vaidya, P. M., & Sung-Mo, K. (1994). Convexity-based algorithms for design centering. *IEEE Transactions on Computer-Aided Design of Integrated Circuits and Systems, 13*(12), 1536–1549.

Shi, Y., & Eberhart, R. (1998). *A modified particle swarm optimizer.* Paper presented at the IEEE International Conference on Evolutionary Computation Proceedings, 1998. IEEE World Congress on Computational Intelligence.

Shi, Y., & Eberhart, R. C. (1999). *Empirical study of particle swarm optimization.* Paper presented at the Congress on Evolutionary Computation, 1999. CEC 99. Proceedings.

Shi, Y., Eberhart, R., & Chen, Y. (1999). Implementation of evolutionary fuzzy systems. *IEEE Transactions on Fuzzy Systems, 7*(2), 109–119.

Xiaohui, H., & Eberhart, R. (2002). *Multiobjective optimization using dynamic neighborhood particle swarm optimization.* Paper presented at the Congress on Evolutionary Computation, 2002. CEC '02. Proceedings.

Yuan, Z., Yang, L., Wu, Y., Liao, L., & Li, G. (2007). *Chaotic particle swarm optimization algorithm for Traveling Salesman Problem.* Paper presented at the 2007 IEEE International Conference on Automation and Logistics.

Zurada, J. M., Lozowski, A., & Malinowski, A. (1997). *Design centering in GaAs IC manufacturing.* Paper presented at the Aerospace Conference, 1997. Proceedings, IEEE.

Trade-offs are not the only constraints affecting the performance and impact of work by living systems. Christina De La Rocha asks us to look at the impact of one of the most fundamental kinds of work on Earth—the work of photosynthesis—from the global perspective provided from biogeochemistry. Starting with the history and basic chemistry of photosynthesis, we see how photosynthesis developed as a work process. We also find that, like many human work processes, it has a choke point, induced by a "working part" that worked just fine when it started but has been unable to change with changing times and conditions. Because this is a crucial part—a critical enzyme—and the changing conditions are the long-term fluctuations in carbon dioxide over the time span of life on Earth, this choke point has come to limit photosynthesis and the work of life on Earth. De La Rocha also takes us through the ways in which carbon makes its way through the biogeochemical cycle, and we can see how the work of photosynthesis and the work of the biosphere affect us all.

4   The Impact of Photosynthetic Work on Earth, Climate, and the Biosphere

Christina De La Rocha

Introduction

The word *work* instantly brings to mind human things—farmers, office workers, factories, and the like—and why not? The work human beings have carried out has raised cities, created civilizations, explored other planets, extended life expectancies, constructed road and communications networks, diminished the world's forested area by half (Bryant, Nielsen, & Tangley, 1997), and started to alter climate. Yet human work is not the bulk of the work carried out on Earth each year. Human work is a small fraction compared to the work carried out on Earth by plants and algae, whose work of collecting energy and creating biomass through photosynthesis underpins the work subsequently carried out by living organisms and has had a greater impact on the environment than anything human beings have yet accomplished.

Like just about all types of work carried out by life, photosynthesis is not carried out as efficiently as it could be; the amount of energy photosynthetic organisms lock up into sugars is only a few percent of the total physically possible. This is not to say, however, that photosynthesis is ineffective. The energy captured through photosynthesis supports a great abundance of the life on Earth. Among these organisms, there is hardly a lipid, carbohydrate, or protein that does not have its origin as an organic compound in photosynthesis—and, within them, not a muscle moves, thought occurs, or cell divides without the expenditure of energy derived from photosynthesis. Even economies and power plants run on photosynthate transformed into coal, gas, and oil.

Photosynthesis, like other types of work carried out by life, impacts the environment. Photosynthetic production of the raw materials from which organisms construct themselves and of the fuel from which biological work is done goes a considerable way toward defining the total amount of life that can exist at any one time on Earth. Photosynthetic production of free oxygen (O_2) has profoundly altered the gas composition of the atmosphere, making possible the evolution of energetic, complex, multicellular animals. Its consumption of carbon dioxide (CO_2) and the sequestration of carbon away from the atmosphere as a side effect of photosynthesis results in an Earth with a climate that is cooler than it would otherwise be.

Carbon, Energy, Work, and the Amount of Life on Earth

If there is any such thing as a physical, tangible currency to the work done by living systems, that currency is carbon. It is the production of organic compounds during photosynthesis that translates energy from the Sun into a form that can be utilized by, stored within, and transferred between organisms. The flux of carbon in and out of the biosphere each year, while not a perfect measure of the amount of work that living systems are accomplishing, is at least an indication of how much solar energy is being consumed to do so.

It is no small amount of carbon that works its way in and out of the biosphere each year. Annually between them, the marine and terrestrial biospheres photosynthetically fix a gross amount of 223×10^{15} grams (g) of carbon (C) into organic matter, a number that is also frequently referred to as 223 petagrams of carbon (Pg C), the units that are used in this chapter. (The same quantity is also referred to in equivalent units as 223 gigatons of carbon (Gt C). A little more than one half of this fixed carbon is converted straight back to CO_2 by the plants and algae themselves, leaving a yearly net input of carbon into the biosphere of 105 Pg C (Field, Behrenfeld, Randerson, & Falkowski, 1998). The respiration of that much organic carbon would yield up 10^{18} kilocalories (kcal) or 4×10^{18} kilojoules (kJ) of energy, the equivalent of a year of 2,000 kcal/day diets for 1.4 trillion people.

When you put this much energy and this much material into a system, many things become possible. On Earth, a multitude of organisms evolved, each with its own approach to obtaining and utilizing organic matter created during photosynthesis. What started out as sunlight and a handful of chemical elements coalesced over time into reefs, jungles, forests, grasslands, marshes, oases, and numerous other interconnected food webs, niches, and ecosystems (see chapter 5).

Virtually all of the organic carbon produced by photosynthesis each year is metabolized by organisms and used to fuel work. Much of the energy goes toward the basic vital processes of cells: transporting materials into, out of, and around the cell; maintaining proton and other electrochemical gradients across membranes (chapter 1); and constructing macromolecules, such as proteins, cholesterol, and DNA. Some of the energy will go toward reproduction, and some of the energy will be used to fuel movement and motion (see chapter 8).

A smaller fraction of the energy stored as photosynthate goes unoxidized. This energy is, in part, tied up in the living mass of the biosphere and, in part, as detrital, nonliving organic matter that has not yet decomposed. The organic carbon in the biosphere is not long lived, being retained there, on average, about 20 years. Organic carbon that settles into sediments may persist considerably longer. Once buried, that material may be lost from the system altogether (or unearthed millennia later by a petroleum company and burned to fuel work).

Because the organic molecules that make up living creatures are also a storehouse of energy, the amount of biomass on Earth reflects the amount of energy available in the biosphere for work. The size of the biosphere is the cumulative amount of photosynthetic work that has been done less the other work living creatures have carried out (and fueled through the respiration of organic matter) since life began. Although the amount of energy coming into the biosphere each year will generally be close to the amount being respired out, photosynthesis has outstripped respiration often enough for the biosphere that we know today to have built up.

The most recent exceptional occurrence of photosynthesis outpacing respiration occurred between about 375 and 300 million year ago and was associated with the rise of vascular land plants (Field et al., 1998) and the subsequent emergence of vertebrates and arthropods onto land. The innovation of woody, structural material (e.g., lignin) that is difficult for bacteria and fungi to decompose and the building up of diverse terrestrial ecosystems resulted in a massive increase in the amount of biomass on Earth. This excess of organic matter production over organic matter degradation sent the oxygen content of the atmosphere soaring (to about 35%, compared with today's 20%) (Berner, 1999) for about 100 million years. This hyperoxia, in turn, made briefly possible the existence of gigantic insects, such as dragonflies with wingspans up to 75 cm and millipedes that were up to 2 meters long, and the increased density of the atmosphere may have facilitated the evolution of wings and flight (Graham, Dudley, Aguilar, & Gans, 1995).

Photosynthesis, Photosynthetic Efficiency, and the Input of Energy to the Biosphere

Work requires energy, and that means that getting work done requires carrying out the initial work of collecting energy. The biosphere has two external energy sources it taps into: photons reaching the surface of the Earth from the Sun and residual heat and reduced chemicals escaping from the interior of the Earth. The overwhelming fraction of energy input to the biosphere comes via oxygenic photosynthesis from the Sun.

Although details of the pigments and electron acceptors differ among the classes of photosynthetic organisms, the complex cellular infrastructure of photosynthesis is similar across taxa whose last common ancestor died billions of years ago (Falkowski et al., 2004). This could be interpreted to mean that life developed the most efficient and reliable means possible for converting solar energy into chemical energy and then stuck with it. More likely, it indicates that, having developed an intricate method for photosynthetic carbon fixation that was head and shoulders above anything else at the time, life got stuck with it. It is too deeply embedded into the basic infrastructure of cells and their highly specific and interwoven biochemical cycles to be substantially modified or replaced at this late stage of the game.

During photosynthesis, pigments like chlorophylls, phycobilins, and carotenoids absorb photons of light of specific wavelengths between 360 and 720 nanometers (nm). These light-harvesting compounds are arranged in a precise manner to transfer energy from photons to a reaction center that picks up electrons from the splitting of water into hydrogen ions (also known as protons or H^+) and O_2. The excitation of the reaction center results in the transport of the high-energy electrons down a chain of electron carriers (figure 4.1) that releases energy from the electrons in small steps.

The energy is used to pump protons from the inside of the chloroplast across a membrane and into a membrane-bound vesicle called a thylakoid. The resulting high concentration of H^+ within the thylakoid relative to the cytoplasm of the chloroplast creates a chemical and electrical gradient across the membrane. This gradient drives H^+ through a proton pump back to the outer side of the membrane (chapter 1). Coupled to the passage of protons through the pump is the addition of phosphate to the molecule adenosine diphosphate (ADP) to produce adenosine triphosphate (ATP), the energy carrier molecule that fuels work in every cell of all living creatures on Earth. At the same time, the electrons that have been liberated through the splitting of water in the light reactions of photosynthesis and have traveled down the chain of electron carriers are picked up by the molecule nicotinamide adenine dinucleotide phosphate ($NADP^+$) and are stored as NADPH. The NADPH and ATP generated in these light reactions of photosynthesis are then used to fuel the fixation of CO_2 into organic matter during the Calvin cycle (figure 4.2).

Every organism undertaking oxygenic photosynthesis, from cyanobacteria in geothermal hot springs to fir trees in boreal forests, fixes carbon from CO_2 into organic matter in a series of reactions known as the Calvin cycle (figure 4.3). Three molecules

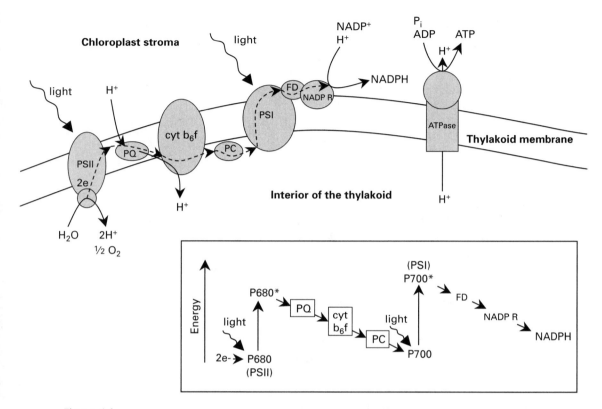

Figure 4.1

The transport of electrons and protons during the light reactions of photosynthesis raises the H⁺ concentration on one side of the thylakoid membrane and lowers it on the other side. The resulting gradient drives protons through the pump, which generates ATP. The inset shows the relative energy of the electrons liberated through the splitting of water following their excitation and transfer down the electron transport chain.

Figure based on figures by Govindjee & Govindjee (1975), Alberts et al. (1989), Campbell and Reece (2002), and Taiz & Zeiger (2006).

of CO_2 are brought into the Calvin cycle when they are captured by an enzyme, known as "Rubisco" (for ribulose bisphosphate carboxylase (RuBP carboxylase)). The function of Rubisco is to attach the three molecules of CO_2 to three molecules of ribulose bisphosphate (RuBP) in a process known as carboxylation. With the input of energy and of hydrogen ions from the breakdown of ATP and NADPH, the carboxylated RuBP molecules are rearranged in a series of steps into six molecules of glyceraldehyde phosphate (GAP). Every sixth molecule of GAP is harvested from the cycle and used in the construction of glucose, generally referred to as the end product of

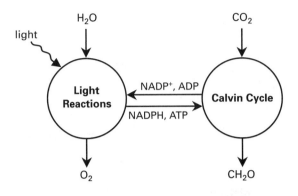

Figure 4.2
An overview of the materials passing between and through the light reactions of photosynthesis and the Calvin cycle. The conversion of CO_2 to sugar during the Calvin cycle consumes ATP and NADPH, converting them back to ADP and $NADP^+$. From these compounds, ATP and NADPH are regenerated during the light reactions of photosynthesis using energy, protons, and electrons derived from light energy and the splitting of water.
Figure based on figures by Alberts et al. (1989) and Campbell and Reece (2002).

photosynthesis. To close the loop, additional ATP is consumed to convert the remaining five GAP molecules back to RuBP. The cycle is then ready to fix the next three molecules of CO_2 into organic matter.

Under ideal conditions, it takes 48 photons of light, 6 molecules of CO_2, 18 molecules of ATP, and 12 molecules of NADPH to produce one molecule of glucose ($C_6H_{12}O_6$). Ignoring losses, such as to photorespiration, roughly 2,880 kcal (12,000 kJ) of energy go into the production of one mole (6.02×10^{23} molecules) of glucose. When this mole of glucose is oxidized back to CO_2, 686 kcal (2,870 kJ) are released to fuel work.

Given the uniformity in the infrastructure of the light reactions of photosynthesis among photosynthetic taxa, and the existence of Rubisco and the Calvin cycle in every oxygenic photosynthesizer, it is tempting to wonder, is this the most efficient way to convert solar energy into the chemical energy of organic matter?

In fact, is it efficient at all? Every year 2.75×10^{24} J of energy reach the Earth's surface from the Sun. After losses to plant respiration and photorespiration, a net 0.15% of this energy is transferred into the biosphere in the form of photosynthate (Archer & Barber, 2004). One cause of this low overall photosynthetic efficiency is that photons are absorbed by things other than photosynthetic organisms. In addition, photosynthesis itself is fairly inefficient. Only 3% to 5% of the solar energy reaching plants and phytoplankton becomes locked up in organic matter.

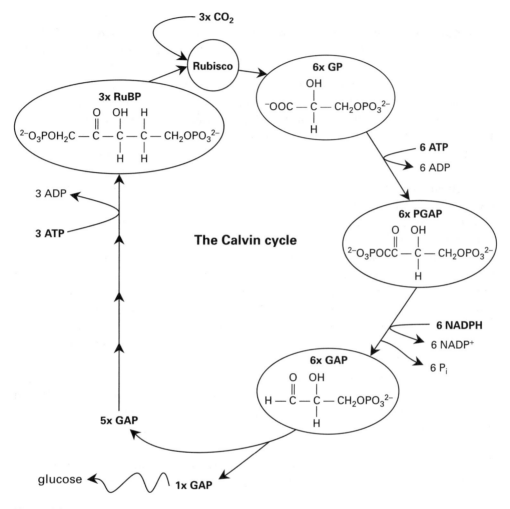

Figure 4.3

The Calvin cycle during which CO_2 is transformed into sugar. First, the enzyme, Rubisco, binds three molecules of CO_2 to three molecules of RuBP to produce six molecules of the 3-carbon organic compound, phosphoglycerate (GP). Through the input of energy and phosphate from the conversion of ATP to ADP, the six GP molecules are converted into six molecules of bisphosphoglycerate (PGAP). The oxidation of NADPH then allows for the restructuring of the six molecules of PGAP into six molecules of GAP. One of every six molecules of GAP is harvested from the cycle and used to construct glucose. The remaining five GAP molecules are converted, in a series of steps, back to three molecules of RuBP.

Figure based on figures by Alberts et al. (1989) and Campbell and Reece (2002).

The 3% to 5% net efficiency of photosynthesis represents energy losses at a number of steps (Archer & Barber, 2004). Nearly one half of the solar energy reaching photosynthetic organisms is lost because the light-collecting compounds of photosynthetic organisms can only utilize light of wavelengths from 360 to 720 nm, what is known as photosynthetically active radiation (PAR). One third of PAR energy is then lost to imperfect absorbance by the chloroplasts of photosynthetic organisms or to absorption by nonphotosynthetic parts of them. One quarter of absorbed PAR energy is lost as the energy of photons of various wavelengths is brought down to the excitation energy at 700 nm by photosystems within chloroplasts. Nearly 70% of the remaining energy is lost in the conversion of excitation energy to the chemical energy of glucose. Plants themselves then use up to half of their gross photosynthetic gain through respiration and photorespiration.

The nonexistence of more efficient mechanisms for some steps in photosynthesis (e.g., the transfer of light energy to the electrochemical energy of ATP and NADPH) makes it difficult to say whether these steps in photosynthesis are being carried out in the most efficient manner possible. With respect to the carbon-fixing reactions of the Calvin cycle, however, it is in fact possible to say that there are more effective ways to operate under the environmental conditions of the present day.

The inception of Rubisco occurred early in Earth's history when CO_2 was considerably more abundant, and there was little atmospheric O_2. At this time, the enzyme's high capacity for CO_2 was advantageous, and its inefficient functioning at low concentrations of CO_2 was irrelevant. At the considerably lower CO_2 concentrations of the modern atmosphere, however, Rubisco cannot operate at its maximal rate (Edwards & Walker, 2004; Raven, Cockell, & De La Rocha, 2008) and may hinder the overall rate of photosynthetic carbon fixation. O_2 also competes with CO_2 for binding sites on Rubisco. At high concentrations of O_2 and low concentrations of CO_2, O_2 wins out. When this process, known as *photorespiration*, happens, Rubisco catalyzes the conversion of organic carbon back to CO_2, and considerable energy is lost.

Photorespiration is enough of a problem that quite a number of photosynthetic organisms have evolved mechanisms to counter the low affinity of Rubisco for CO_2. Most notably, *C_4 plants*, such as maize and many grasses, have adopted the use of an enzyme, phosphoenolpyruvate carboxylase (PEP carboxylase), with a higher affinity for CO_2, to concentrate CO_2 for the benefit of Rubisco. PEP carboxylase is used to take up atmospheric CO_2 and combine it with phosphoenolpyruvate (PEP) to produce a C_4 compound that is later converted back to CO_2 within chloroplasts. The effect of this process is to keep Rubisco bathed in a relatively high CO_2 environment.

The operation of an extra biochemical pathway and the construction of an extra carbon-fixing enzyme require C_4 plants to expend extra energy, resources, and infrastructure. C_4 photosynthesis is nonetheless more efficient and effective than the traditional C_3 photosynthesis under certain conditions. This is most true on hot, sunny days when land plants close their stomata to reduce evaporative water losses. Stomata are also used for gas exchange. Because photosynthesis continues even when stomata are closed, CO_2 concentrations within photosynthetic cells decline and O_2 concentrations rise, increasing the incidence of photorespiration. Under these conditions, C_3 plants may photorespire away 30% to 40% of the gross yield of photosynthesis (Archer & Barber, 2004). By reducing the occurrence of photorespiration, C_4 plants can be considerably more efficient at converting energy of photons absorbed to glucose than the 3% to 5% of C_3 plants (Archer & Barber, 2004).

From an evolutionary perspective, it is interesting that despite PEP carboxylase's carbon-fixing abilities, higher affinity for CO_2, and clear advantage over Rubisco, PEP carboxylase has not replaced the use of Rubisco in plants. Because PEP carboxylase can carboxylate PEP but not RuBP, replacing Rubisco with PEP carboxylase would require replacing the entire Calvin cycle (figure 4.3). Alternatively, the entire Calvin cycle would have to be replaced with a cycle that could construct GAP using NADPH, ATP, and carboxylated PEP and regenerate PEP as part of the process. The evolution of an entirely new biochemical cycle to create sugar from energy and CO_2 is too tall an order in a world in which habitat space is already filled up with photo-synthetic organisms, even if they are not perfectly efficient. The best that photosynthetic organisms can do to improve the efficiency of their photosynthetic apparatuses is to tinker within the context of the infrastructure and enzymes that have already evolved.

Photosynthesis, the Composition of the Atmosphere, and Implications for the Energetics of Life

The work of the biosphere can have far-reaching environmental consequences that have great impact on the way organisms in the biosphere do work. Oxygenic photosynthesis is the ultimate example of this.

For the first few billion years of Earth's history, little more than trace amounts of free oxygen occurred in the atmosphere. Respiration was carried out anaerobically, using chemicals like sulfate instead of oxygen. The primary production of organic matter by living systems was carried out, both chemosynthetically and photosynthetically, without generating O_2 (Holland, 2003).

Sometime before 2.7 billion years ago, oxygenic photosynthesis developed (Xiong, Inoue, Nakahara, & Bauer, 2000), and conditions on Earth began to change (Scott et al., 2008). First, oxygenic photosynthesis increased the flow of energy into the biosphere, greatly increasing its capacity for work. This was a direct result of shifting from a limited pool of reduced chemicals to an effectively unlimited supply of water as the source of protons and electrons for chemiosmosis (Dismukes et al., 2001). As a result, rates of primary production increased, as did the overall mass of living creatures. As the biomass of oxygenic photosynthesizers increased, and global rates of oxygenic photosynthesis increased further, O_2 was produced at a greater and greater rate. Intermittently, the overwhelmingly anoxic waters of the ocean and air of the atmosphere became strongly oxygenated as the atmosphere went from containing only trace amounts of O_2 to containing the 21% O_2 of the atmosphere today (Holland, 2003; Scott et al., 2008).

One of the first effects that the addition of O_2 to the atmosphere had on Earth was the establishment of the ozone layer. The chemical reactions involved still occur today. Interactions between short-wavelength solar ultraviolet (UV) radiation and O_2 in the stratosphere create free oxygen atoms (O) that then combine with O_2 to produce ozone (O_3). When this resulting stratospheric O_3 absorbs longer wavelength UV radiation, the energy it absorbs photodissociates, or splits, O_3 back into O_2 and O. It is this O_3-destroying reaction that shields the surface of the Earth from solar UV radiation, making shallow waters and the land surface habitable.

The oxygenation of the ocean and atmosphere also paved the way for aerobic respiration and the subsequent evolution of complex, multicellular (and eventually large) animals.

Because oxygen is a far more powerful electron acceptor than those used in anaerobic respiration, aerobic respiration extracts nearly 20 times more energy for work from glucose than anaerobic means, such as sulfate reduction or methanogenesis (Dismukes et al., 2001). The high demand for energy of complex, multicellular organisms seems to require this more efficient process for the extraction of energy from organic carbon. With the exception of some anaerobic parasitic nematodes, the forms of life that are entirely supported through anaerobic respiration are unicellular. Likewise, from the fossil record, it appears that the emergence and unprecedentedly rapid evolution of complex, multicellular animals during the Cambrian Explosion (around 550 million years ago) required the widespread availability of oxygen for respiration (McFadden et al., 2008).

These two evolutionary innovations, oxygenic photosynthesis and aerobic respiration, increased the number of ways life could do the work of building up and exploiting stores of energy and, in so doing, vastly expanded the volumetric extent of the Earth's biosphere. As time has passed, the accumulation of innovations in the accomplishment of work has resulted in the colonization, from that first restricted locality where life first took hold, of the entire ocean and the deep subsurface, to rivers, streams, and lakes, and finally to land, and, to some degree, the sky. Today we live in a world in which aerobic respiration both dominates and coexists with a considerable diversity in the means of extracting energy from organic matter to fuel work (e.g., sulfate reduction, methanogenesis, denitrification, and fermentation); furthermore, these various processes may occur only millimeters apart (e.g., in wetland sediments), separated by gradients in oxygen concentrations.

Primary Production and the Greenhouse Effect

Not only has the work of photosynthesis revolutionized Earth's surface conditions and made it possible for organisms to fuel work through aerobic respiration, it continues to have a significant effect on global climate through its impact on concentrations of CO_2 in the atmosphere.

In addition to being raw material for photosynthesis, CO_2 is a greenhouse gas. It absorbs strongly in the infrared, slowing the radiation of heat out to space, causing Earth's temperature to rise. Given Earth's distance from the Sun, if it had no greenhouse gases, its average surface temperature would be about −20°C (Broecker, 1985), roughly the same as the moon's. Instead, Earth's surface temperatures average 15°C, warm enough for palm trees. The bulk of this greenhouse warming is carried out by water vapor, even though the overall effect of water vapor as a greenhouse gas is moderated by its short-term variability and its ability to condense to form clouds, which have a cooling effect. It is, however, changes in CO_2 that occur over longer timescales that impart both finer scale control and longer term stability to Earth's climate.

Most of the rapidly reacting carbon in the surface Earth–ocean–atmosphere system resides in dissolved form in the ocean (figure 4.4) (Sundquist, 1993; Prentice et al., 2001). Ocean waters contain 37,000 Pg C as dissolved inorganic carbon (DIC), which consists of dissolved CO_2, carbonic acid, bicarbonate ion, and carbonate ion. By comparison, the terrestrial biosphere, including detritus in soils, holds only about 2,000 Pg C. This leaves only a smaller amount of carbon (750 Pg C) to reside in the atmosphere in the form of CO_2.

The exact amount of CO_2 in the atmosphere on geologically short timescales depends on the balance that is struck between surface ocean waters and the atmosphere. These waters are in direct contact with the atmosphere and remain at the surface long enough for a rough equilibrium to be reached between them with respect to CO_2. The anthropogenic CO_2 emitted to the atmosphere is thus invading the ocean, increasing the CO_2 content of surface ocean waters and driving the observed rapid acidification of these waters (Wootton, Pfister, & Forester, 2008). Likewise, any process that removes CO_2 from surface waters subsequently lowers atmospheric concentrations of CO_2.

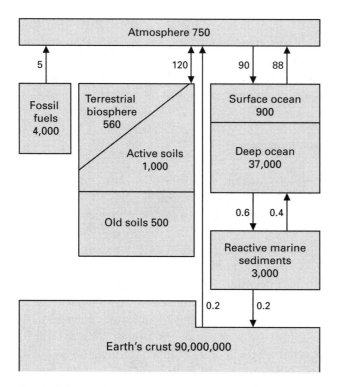

Figure 4.4
The reservoirs and fluxes of the global carbon cycle (redrawn from Sundquist (1993) with flux values updated according to Prentice et al. (2001)). Reservoir sizes are given in petagrams of carbon and fluxes are given in petagrams of carbon per year. The forms of carbon covered by the diagram include CO_2, dissolved organic carbon or DIC (H_2CO_3, HCO_3^-, CO_3^{2-}), particulate organic carbon, and calcium carbonate ($CaCO_3$).
From Sundquist, E. T. (1993). The global carbon dioxide budget. *Science*, *259*, 934–941. Reprinted with permission from AAAS.

The interesting twist for climate is that the ocean can be partitioned into two main sections in terms of its DIC concentration. There are those surface waters, the upper approximately 100 meters of the water column, where DIC concentrations are at their lowest. Underneath the surface waters, there is a density barrier, the thermocline, due to the lower temperatures and higher salinities of deeper waters. Below the thermocline, concentrations of DIC are about 10% higher than in surface waters; deep water brought to the surface emits significant quantities of CO_2 to the atmosphere.

The thermocline acts as a cap limiting the contact of the CO_2-rich deep waters with the atmosphere. The deep waters become enriched in CO_2 via a series of processes known as the *biological pump* (e.g., De La Rocha, 2007). This transfer of CO_2 out of the atmosphere and into the deep ocean begins with the work of phytoplankton and has an impact on global temperatures.

The production of organic matter from CO_2 and sunlight during marine photosynthesis results in a lowering of the concentration of CO_2 in surface waters. Most of this organic matter works its way through marine food webs, providing the energy and biomass for life in the ocean, and is quickly respired back into CO_2. A small fraction, however, survives long enough before decomposition to sink—as aggregated clusters of phytoplankton, as the discarded, sticky, weblike feeding structures of certain zooplankton, and as fecal pellets of zooplankton—into the deep sea and sediments (Alldredge and Silver, 1988; De La Rocha, 2007). The decomposition of this material at depth adds CO_2 to deep waters, sequestering this CO_2 away from the atmosphere for a few hundred to about a thousand years. The amount of carbon that the biological pump transports to the deep ocean relative to the amount of CO_2 that comes to the surface during the upwelling of subthermocline waters controls the concentration of CO_2 in the atmosphere over timescales of thousands of years.

Paleoenvironmental records show that atmospheric CO_2 and mean surface temperature are strongly correlated between glacial and interglacial periods over at least the last 630,000 years (Cuffey & Vimeux, 2001; Siegenthaler et al., 2005). A shift in CO_2, for example, from the 190 µatm (µatm is equivalent to 1 part per million by volume) of the last glacial maximum to the 280 µatm of the (preindustrial) Holocene interglacial may mean the difference between sheets of ice kilometers high on northern Europe and most of Canada and the warmth of the present day.

Numerous physical, chemical, biological, and astronomical factors working in concert drive the glacial–interglacial shifts in CO_2 in ways that we do not fully understand. However, the ocean's biological pump has a considerable role to play in setting these values (Sigman and Boyle, 2000). Were all life to die in the ocean and the DIC gradient between surface and deep waters to erode over time, atmospheric CO_2 concentrations would rise, eventually stabilizing at around 415 µatm (Broecker, 1982), driving up global temperatures. Alternatively, if all the nutrients in ocean surface waters were completely utilized in support of photosynthesis (in some regions of the ocean, nutrients do not get completely used up each year), the biological pump could lower atmospheric CO_2 concentrations well below the 190 µatm of glacial periods.

Although CO_2 may drive changes in global temperature over these geologically short timescales, over longer timescales it is the greenhouse gas responsible for the surprising stability of the Earth's climate (Berner, 1998). The chemical weathering that breaks down rocks into the salts that make their way to the ocean also converts CO_2 into carbonate ions and bicarbonate ions that also eventually end up in the sea. Such weathering reactions vary with temperature, such that when temperatures are high, chemical weathering rates are faster, removing CO_2 from the atmosphere at a higher rate and consequently lowering global temperatures. As global temperatures drop, so do the rates of chemical weathering and CO_2 consumption, allowing CO_2 concentrations to build back up again. Although this is a slow process, reaching a balance only on timescales of millions of years, it has held global average temperatures on Earth within a fairly temperate range continuously over the 4 billion years it took to evolve complex, multicellular, intelligent animals.

Complicating this balance, on the other hand, is the work of life itself. It is highly likely that appearance of vascular land plants several hundred million years ago changed the relationship between temperature and weathering rates by increasing the rates at which minerals could be weathered at any given temperature. The two things that land plants do that could have led to such an increase in weathering rates are to break up rocks with their roots and to exhale CO_2 from their roots, allowing high concentrations of CO_2 to build up in areas of the soil profile where active weathering takes place. The activities of land plants should thus result in a lowering of the set point temperature around which the long-term CO_2 cycle works to stabilize climate (Berner, 1998).

In both the case of the biological pump in the ocean and of the impact of land plants on rock weathering, an unexpected effect of the photosynthetic input of energy into the biosphere is that the climate of Earth is cooler than it would be in the absence of life.

Concluding Remarks

The size of the biosphere on Earth and the amount of work it does ultimately depends on the input of energy from oxygenic photosynthesis. Although oxygenic photosynthesis provides the biosphere with a net input of 4×10^{18} kJ of energy each year, in the form of 105 Pg C fixed into organic matter, oxygenic photosynthesis is inefficient at converting solar energy into chemical energy. There are undoubtedly ways to alter photosynthetic apparatuses to increase the efficiency of photosynthesis, but the remarkable constancy of these apparatuses across all oxygenic photosynthetic organisms suggests that it is not possible to make major changes to the process of photosynthesis through evolution in a world with habitats full of successfully photosynthesizing organisms.

The innovation of the work of oxygenic photosynthesis changed the way life on Earth worked by vastly increasing the amount of energy that could be input into the biosphere and by oxygenating the atmosphere, making aerobic respiration (and, thus, complex, multicellular animal life) possible. The work of photosynthetic organisms has other far-reaching impacts, such as the sequestration of carbon in the deep ocean and sediments and the enhancement of rock-weathering rates, two processes that result in cooler global temperatures.

References

Alberts, B., Bray, D., Lewis, J., Raff, M., Roberts, K., & Watson, J. D. (1989). *Molecular biology of the cell* (2nd ed.). New York: Garland Publishing.

Alldredge, A. L., & Silver, M. W. (1988). The characteristics, dynamics, and significance of marine snow. *Progress in Oceanography, 20*, 41–82.

Archer, M. D., & Barber, J. (2004). Photosynthesis and photoconversion. In M. D. Archer & J. Barber (Eds.), *Molecular to global photosynthesis* (pp. 1–41). London: Imperial College Press.

Berner, R. A. (1998). The carbon cycle and CO_2 over Phanerozoic time: The role of land plants. *Philosophical Transactions of the Royal Society of London. Series B, Biological Sciences, 353*, 75–82.

Berner, R. A. (1999). Atmospheric oxygen over Phanerozoic time. *Proceedings of the National Academy of Sciences of the United States of America, 96*, 10955–10957.

Broecker, W. S. (1982). Ocean chemistry during glacial time. *Geochimica et Cosmochimica Acta, 46*, 1689–1705.

Broecker, W. S. (1985). *How to build a habitable planet.* Palisades, NY: Eldigio Press.

Bryant, D., Nielsen, D., & Tangley, L. (1997). Last frontier forests: Ecosystems and economies on the edge. Washington, D.C.: World Resources Institute. Retrieved from http://www.wri.org/publication/last-frontier-forests.

Campbell, N. A., & Reece, J. B. (2002). *Biology* (6th ed.). San Francisco: Benjamin Cummings.

Cuffey, K. M., & Vimeux, F. (2001). Covariation of carbon dioxide and temperature from the Vostok ice core after deuterium-excess correction. *Nature, 412*, 523–527.

De La Rocha, C. L. (2007). The biological pump. In H. D. Holland & K. Turekian (Series Eds.), *Treatise on geochemistry update 1: Vol. 6: The oceans and marine geochemistry*, H. Elderfield (Ed.), (pp. 1–29). Oxford: Elsevier Pergamon. doi:0.1016/B0–08–043751–6/06107–7.

Dismukes, G. C., Klimov, V. V., Baranov, S. V., Kozlov, Y. N., DasGupta, J., & Tyryshkin, A. (2001). The origin of atmospheric oxygen on Earth: The innovation of oxygenic photosynthesis. *Proceedings of the National Academy of Sciences of the United States of America, 98*, 2170–2175.

Edwards, G. E., & Walker, D. A. (2004). Photosynthetic carbon assimilation. In M. D. Archer & J. Barber (Eds.), *Molecular to global photosynthesis* (pp. 189–220). London: Imperial College Press.

Falkowski, P. G., Katz, M. E., Knoll, A. H., Quigg, A., Raven, J. A., Schofield, O., & Taylor, F. J. R. (2004). The evolution of modern eukaryotic phytoplankton. *Science, 305*, 354–360.

Field, C. B., Behrenfeld, M. J., Randerson, J. T., & Falkowski, P. (1998). Primary production of the biosphere: Integrating terrestrial and oceanic components. *Science, 281*, 237–240.

Govindjee & Govindjee, R. (1975). Introduction to photosynthesis. In Govindjee (Ed.), *Bioenergetics of photosynthesis* (pp. 1–50). New York: Academic Press.

Graham, J. B., Dudley, R., Aguilar, N. M., & Gans, C. (1995). Implications of the late Palaeozoic oxygen pulse for physiology and evolution. *Nature, 375*, 117–120.

Holland, H. D. (2003). The geologic history of seawater. In H. D. Holland & K. Turekian (Series Eds.), *Treatise on geochemistry. Vol. 6: The oceans and marine geochemistry*, H. Elderfield (Ed.), (pp. 583–625). Oxford: Elsevier-Pergamon.

McFadden, K. A., Huang, J., Chu, X., Jiang, G., Kaufman, A. J., Zhou, C., ... Xiao, S. (2008). Pulsed oxidation and biological evolution in the Ediacaran Doushantuo Formation. *Proceedings of the National Academy of Sciences of the United States of America, 105*, 3197–3202.

Prentice, I. C., Farquhar, G. D., Fasham, M. J. R., Goulden, M. L., Heimann, M., Jaramillo,V. J., ... Yool, A. (2001). The carbon cycle and atmospheric carbon dioxide. In J. T. Houghton, Y. Ding, D. J. Griggs, M. Noguer, P. J. van der Linden, X. Dai, K. Maskell, & C. A. Johnson (Eds.), *Climate change 2001: The scientific basis. Contribution of Working Group I to the Third Assessment Report of the Intergovernmental Panel on Climate Change* (pp. 183–237). Cambridge, England: Cambridge University Press.

Raven, J. A., Cockell, C. S., & De La Rocha, C. L. (2008). The evolution of inorganic carbon concentrating mechanisms in photosynthesis. *Philosophical Transactions of the Royal Society B, 363*, 2641–2650.

Scott, C., Lyons, T. W., Bekker, A., Shen, Y., Poulton, S. W., Chu, X., & Anbar, A. D. (2008). Tracing the stepwise oxygenation of the Proterozoic ocean. *Nature, 452*, 456–459.

Siegenthaler, U., Stocker, T. F., Monnin, E., Luthi, D., Schwander, J., Stauffer, B., ... Jouzel, J. (2005). Stable carbon cycle-climate relationship during the late Pleistocene. *Science, 310*, 1313–1317.

Sigman, D. M., & Boyle, E. A. (2000). Glacial/interglacial variations in atmospheric carbon dioxide. *Nature, 407*, 859–869.

Sundquist, E. T. (1993). The global carbon dioxide budget. *Science, 259*, 934–941.

Taiz, L., & Zeiger, E. (2006). *Plant physiology* (4th ed.). Sunderland, MA: Sinauer Associates, Inc.

Wootton, J. T., Pfister, C. A., & Forester, J. D. (2008). Dynamic patterns and ecological impacts of declining ocean pH in a high-resolution multi-year dataset. *Proceedings of the National Academy of Sciences of the United States of America, 105*, 18848–18853.

Xiong, J., Inoue, K., Nakahara, M., & Bauer, C. E. (2000). Molecular evidence for the early evolution of photosynthesis. *Science, 289*, 1724–1730.

The sense that the activities of organisms affect the way that organisms evolve and work on Earth is not a new one for Kevin Laland and Gillian Brown. First, they describe the relationship between the evolutionary theory of niche construction *and the study of work. Niche construction holds that organisms are not merely selected for niches by natural selection, but that they actively shape their niches and those of other organisms through their own activity—as De La Rocha has demonstrated in chapter 4. The work of an organism is, therefore, a vital force in the ongoing process of evolution. Humans are, in this way, the paramount* niche constructors, *though by no means the only ones—all organisms niche construct. Next, Laland and Brown explore evolutionary aspects of human work by applying the theories of human behavioral ecology to the study of work and the modern workplace. In this way, they synthesize their own research interests in evolutionary biology and animal behavior. They develop connections between the thermodynamic and everyday senses of* work *and explore connections with concepts of energetic models of fitness and energetic trade-offs.*

5   Niche Construction and Human Behavioral Ecology: Tools for Understanding Work

Kevin Laland and Gillian Brown

What is Work?

Living organisms are far-from-equilibrium (strongly out-of-equilibrium) systems relative to their physical or abiotic surroundings. They can only survive and maintain their far-from-equilibrium status by constantly exchanging energy and matter with their environments. Organisms feed on molecules rich in free energy and, in the process, generate outputs largely in the form of molecules that are poor in free energy. The energy harvested is used to do work. Such work is necessary to allow organisms to produce and maintain order, both inside their bodies and in their external environments. Thus, in order to survive, organisms must act on their environments and, by doing so, change them. One consequence of this imperative is that all living organisms must engage in *niche construction*—that is, they must modify their environment to some degree. Human-constructed artifacts are testament to the extraordinary lengths humans take niche construction, but the process is present across all forms of life. This chapter explores how the concept of niche construction helps us to understand the relationships between organisms and their environments and provides examples of how the transfer of energy through niche-constructing activities influences the evolutionary process.

First, we need to examine a definition of *work*. Work in the thermodynamic sense is only done when energy flows (Turner, 2000). Hence, for biologists, work is defined as the processes that organisms engage in that allow them to exchange energy with their environments, to channel energy through their bodies, and to create orderliness in their world. Such work is a necessary condition for organisms to survive. What, if anything, is the relationship between this definition of *work* and uses of the term that encapsulate human vocational activities in the workplace, such as teacher, construction site worker, or information technology consultant? At first glance, there may not appear to be a connection between these two uses of the term. One of the interesting things about the concept of niche construction is that it bridges these two very different senses of work. Through niche construction theory, it can be shown that there is a direct relationship, although *work* in the thermodynamic sense and *work* in the vocational sense are certainly not one and the same. This chapter introduces the fields of niche construction and human behavioral ecology; critiques the idea that human beings are adapted to past but not current environments; and discusses the implications of this line of reasoning for the study of human work, returning to the connections between the energetic and vocational senses of "work."

An Introduction to Niche-Construction Theory

Niche-construction theory is a relatively recent development in evolutionary biology that has gathered momentum over time (Laland, Odling-Smee, & Feldman, 1996, 1999, 2001, 2004; Lewontin, 1982, 1983; Odling-Smee, 1988; Odling-Smee, Laland, & Feldman, 1996, 2003). Advocates of the niche-construction perspective are concerned with the nature of the process of evolution, with the causal basis of the organism–environment match and its inherent symmetries, and with the active role that organisms play in driving evolutionary and coevolutionary events. The perspective seeks to explain the adaptive complementarity of organisms and their environment in terms of a dynamic, reciprocal interaction between the processes of natural selection and niche construction.

Niche construction is the process whereby organisms, through both the metabolic and physical work that they do, modify their own niches and often each other's (Odling-Smee et al., 2003). Historically, the shaping of the environment through the niche-constructing activity of the organism has been overlooked in evolutionary theory, with environments being considered the source of selection and the determiner of the features of living creatures. According to Williams (1992, p. 484): "Adaptation is always asymmetrical; organisms adapt to their environment, never vice versa."

However, organisms evidently do bring about changes in environments. Numerous animals manufacture nests, burrows, holes, webs, and pupal cases; plants change levels of atmospheric gases and modify nutrient cycles; fungi decompose organic matter; and bacteria engage in decomposition and nutrient fixation (see Odling-Smee et al. 2003, for a review of this literature). Organisms also deplete and destroy important components of their world, and the term *niche construction* refers to both positive and negative ramifications of organisms' activities.

Niche construction goes beyond the building of environmental components by organisms. Organisms also regulate their constructions in order to damp out variability in environmental conditions. In generation after generation, beavers, earthworms, ants, and countless other animals build complex structures and artifacts, regulate temperature and humidity inside them, control nutrient cycling and chemical ratios and concentrations around them, and, in the process, construct, maintain, and defend benign and apposite nursery environments for their offspring. For instance, the nest of the termite *Cephalotermes rectangularis*, by virtue of having a thick outer wall permeated with a labyrinth of fine galleries, is ideal for protecting the occupants from external extremes of temperature (Hansell, 1984). Investment of energy in the construction of the mound results in enhanced survivorship of the termites under external environmental conditions that would be stressful, if not lethal, to the developing offspring.

From the niche-construction perspective, evolution is based on networks of causation and feedback: The work of organisms drives environmental change, and organism-modified environments subsequently select adaptive characteristics of organisms (Lewontin, 1983; Odling-Smee et al., 2003). Standard evolutionary theory, on the other hand, models the evolutionary consequences of niche construction solely in terms of fitness payoffs to the genes expressed in niche construction. For instance, the only feedback from a beaver's dam widely considered to be evolutionarily significant is that which affects the fitness of genes that are expressed in building this extended phenotype, relative to their alleles (e.g., Dawkins, 1982).

The conventional approach, however, misses part of the causal story. When a beaver builds a dam and lodge, creating a lake and influencing river flow, it not only affects the propagation of dam-building genes, but it dramatically changes its local environment (Naiman, Johnston, & Kelley, 1988, p. 753):

These [beaver] activities ... modify nutrient cycling and decomposition dynamics, modify the structure and dynamics of the riparian zone, influence the character of water and materials transported downstream, and ultimately influence plant and community composition and diversity.

It follows that beaver dam-building must also transform the selection acting on a host of other traits of beavers, in turn influencing subsequent beaver evolution. The agency of beavers in constructing these modified selection pressures and thereby acting as drivers and codirectors of their own evolution—not to mention that of other species—

too often goes unrecognized. Advocates of the niche-construction perspective feel that the role of the organism as part cause of the evolutionary process is being short-changed by the conventional approach. The concern is exacerbated in the cases of species with large brains and complex behavior, particularly in cases where niche construction is not well described as "caused by genes." The concern is particularly germane to hominin evolution and to evolutionary accounts of human behavior.

The conventional view also neglects the fact that some organism-driven changes in the environment persist as a legacy to modify selection on subsequent generations as an *ecological inheritance* (Odling-Smee, 1988). For example, the modified selection pressures will remain in the beaver's environment as long as the dam, lake, or lodge remain; given that dams are frequently maintained by families of beavers for decades (Naiman et al., 1988), this could be considerably longer than the lifetime of an individual beaver. Offspring inherit two legacies from their ancestors: genes and a modified selective environment.

The conventional view also neglects the fact that characters acquired through learning can play an evolutionary role by influencing niche construction. For instance, the Galapagos woodpecker finch learns to use a cactus spine or similar implement to grub for insects in the bark of trees (Tebbich, Taborsky, Febl, & Blomqvist, 2001). This behavior is not guaranteed by the presence of naturally selected genes, but it has modified the selection acting on these birds to favor a beak optimal for tool use rather than for pecking wood.

The role of acquired characters becomes of particular significance among vertebrate evolution, as a result of their flexible, brain-based learning. There is already considerable interest among evolutionary biologists in the roles that song learning, filial imprinting, habitat imprinting, cultural transmission, and other forms of learning play in such evolutionary processes as speciation, the evolution of adaptive specializations, adaptive radiations, the colonization of new habitats, brood parasitism, and sexual selection in vertebrates. The significance of acquired characters to evolutionary processes becomes amplified with stable transgenerational culture. It is now widely believed that such characters were probably extremely important to hominid evolution (Richerson & Boyd, 2005; Laland & Brown, 2006).

The Coevolution of Dairy Farming and Lactose Absorption: A Case Study

A good example illustrating the evolutionary credentials of niche construction is the coevolution of dairy farming and lactose absorption. While dairy farming is an old, widespread, but not universal form of human work, adult humans vary considerably in their ability to digest milk and consume dairy products without sickness due to differences in digestive physiology (Durham, 1991). Consuming dairy products makes the majority of adult humans ill, because the activity level of the enzyme lactase in their bodies is insufficient to break down the lactose in dairy products; whether they can do so depends on whether individuals possess the appropriate allele. A strong correlation exists between the incidence of the gene for lactose absorption and a history of dairy farming in populations (Ulijaszek & Strickland, 1993). Dairy products are a useful source of energy and protein, and this has led to the hypothesis that dairy farming created the selection pressures that led alleles for lactose absorption to become common in pastoralist communities.

Following earlier work by Aoki (1986), Feldman and Cavalli-Sforza (1989) used mathematical population-genetic models to investigate this possibility. Their analysis showed that whether an allele allowing adult milk digestion achieved a high frequency depended on the fidelity of cultural transmission; that is, on the probability that the children of dairy-product users themselves became milk consumers. If this probability was high, then a significant fitness advantage to the genetic capacity for lactose absorption resulted in the selection of the absorption allele to high frequency within the 6,000 years, or 300 generations, available from the advent of dairy farming to the present day. The analysis accounts for the spread of lactose absorption in those pastoralist societies that use dairy products widely and consistently. Recently, Feldman and Cavalli-Sforza's conclusions have received support from a comparative analysis by Holden and Mace (1997), which concluded that dairy farming spread before the genes for lactose absorption.

The widespread adoption of dairy farming in Europe, Africa, and America—and the associated breeding of cows; development of infrastructure, methodologies, and technologies; and creation of pastureland—is a set of human niche-constructing activities that result from human cultural processes. There are no "genes for" the production of milk, cheese, yogurt, and whey from the milk of domesticated animals in the sense described by Dawkins (1976). While the dairy farmer's genes are expressed as the cows are milked, this is of no more significance than the fact that his or her muscles are flexed or neurons are firing. Genes do not constitute the appropriate level of analysis to explain why individuals in some societies farm cattle and others do not. This is a cultural phenomenon, and explanations at the genetic level would be both inept and overly reductionist. The English do not manufacture stilton, the French brie, and the Dutch edam because of differences in their genes. Neither is dairy farming an adaptation in the sense described by Williams (1966) but rather an adaptive cultural practice. Niche construction is not reducible to prior natural selection; yet, as described above, this work activity has generated selection that favors genes for lactose absorption. Natural selection does not cause this particular niche construction, and neither does drift, mutation, or any of the other established causal processes in evolution; yet niche construction has had clear evolutionary consequences.

At first, it seems that it is the uniquely human capacity for culture that allows for directed, selection-generating niche-constructing activity (i.e., work) that is both stably inherited and not well specified by genes. This would suggest that niche construction that is directed, in part, by human cultural and developmental processes and that is intimately linked with the work that human beings do has an impact on human evolution. But, in fact, such cultural processes are not restricted to humans. To the extent that the manufacture and use of tools are indicative of the possession of culture in all species in the genus *Homo*, cultural niche construction should be recognized as a major factor in hominin evolution. If primatologists are correct in their claim that geographic variation in tool use reflects cultural processes (Whiten et al., 1999), work-driven niche construction may be a factor in primate evolution. In sum, there are likely to be many species for which cultural niche construction plays an evolutionary role, and a multitude of other species in which noncultural niche construction influences evolution (Odling-Smee et al., 2003).

Niche Construction, Human Behavioral Ecology, and the Adaptiveness of Human Behavior

One benefit of niche-construction theory is that it facilitates the use of evolutionary arguments to interpret contemporary human behavior. Here, we consider how a niche-construction perspective affects the standing of the research philosophy of a contemporary evolutionary approach to the study of human behavior; namely, *human behavioral ecology*. An extended account of this argument is given by Laland and Brown (2006).

The human behavioral ecology approach emerged in the late 1970s in the aftermath of the human sociobiology debate (Laland & Brown, 2002). Human behavioral ecology is an approach to understanding human behavior in which anthropologists apply the methods of animal behavioral ecology to human populations. Researchers in the field of animal behavioral ecology ask questions about the survival value, or function, of behavior and begin with the assumption that individual animals behave in a manner that maximizes their reproductive success (Krebs & Davies, 1993). Behavioral ecologists predict that an animal's behavior (including foraging behavior, mate choice, fighting strategies, and parental investment strategies) will be optimal for the specific environment in which the animal lives. Human behavioral ecologists take a similar perspective and explore the extent to which the behavioral differences observed between human groups are responses to particular environments. Their aims are to determine how ecological and social factors affect behavioral variability within and between populations and to predict patterns of behavior using optimality and fitness-maximization models. A key assumption of human behavioral ecology is that human beings are able to alter their behavior flexibly in response to environmental conditions in a manner that optimizes their lifetime reproductive success.

A second feature of the approach is the testing of hypotheses derived from formal or mathematical evolutionary theory. Usually the predictions of the models are tested with data gathered on small communities in remote regions of the world, such as the Ache of Paraguay or the Kipsigis of Kenya. For example, Monique Borgerhoff Mulder (1990) studied the marriage practices in the Kipsigis to investigate whether the circumstances under which women will marry an already married man can be predicted with a mathematical model (the polygyny threshold model) that works well for other animal species. The model made effective predictions.

Critics of this approach have questioned whether human behavior will remain adaptive in modern industrialized societies, which are a recent phenomenon and are physically, technologically, and socially very different from the environments of our ancestors (Symons, 1987; Tooby & Cosmides, 1990). Such critics include evolutionary psychologists, who argue that, over the last 2 million years, our ancestors have spent most of their existence hunting and gathering for a living in small groups in Africa (Tooby & Cosmides, 1990). They suggest that a history of selection will have fashioned human minds to be adapted to this ancestral world of the Pleistocene and not its modern counterpart. To quote Cosmides and Tooby (1987, pp. 280–281):

The recognition that adaptive specializations have been shaped by the statistical features of ancestral environments is especially important in the study of human behavior … Human psychological mechanisms should be adapted to those environments, not necessarily to the twentieth-century industrialized world.

Because evolution is a response to changed selection pressures and that response cannot be instantaneous, all organisms must experience some "adaptive lag." In this respect, humans are not unique. However, leading evolutionary psychologists believe that the adaptive lag for humans is atypically large, because human cultural processes have changed human environments so extensively and so quickly. The idea that modern humans experience a large discordance in their selective environments compared with those to which they are adapted is termed the *adaptive-lag hypothesis*.

Human behavioral ecologists typically respond to the putative problem of adaptive lag by stressing the flexibility of human behavior, which, they claim, allows humans to accommodate themselves to a wide range of environmental circumstances (Smith, Borgerhoff Mulder, & Hill, 2001). Yet even the most adaptable of creatures will experience limits to the environmental conditions in which it can flourish, outside of which it is unable to behave adaptively. Is it too much to expect humans to behave adaptively in modern industrialized worlds? The fact that human behavioral ecologists almost exclusively study people living in preindustrial societies reinforces the view that the adaptive-lag hypothesis may be correct and that modern postindustrial societies may be too different from ancestral human selective environments for humans to behave adaptively. Indeed, leading human behavioral ecologists have indicated that behavioral ecological theory is more likely to be predictive in small-scale, preindustrial societies than in contemporary western societies and have stated they choose to study small-scale societies for this reason (Laland & Brown, 2002).

However, the adaptive-lag problem disappears when one considers human evolution and human work from a niche-construction perspective. Those human behavioral ecologists who maintain human behavior should be adaptive in all human environments are correct, albeit for different reasons than are generally given.

Human niche construction is typically adaptive, at least in the short term, as described in the following sections. Human-constructed environments may appear very different from ancestral ones, but this apparent difference is in important respects illusory. Postindustrial environments, including the modern industrial workplace, share hidden commonalities with the naturalistic environments experienced by our ancestors. Human societies will typically be characterized by less, rather than more, adaptive lag than is typical for other animals.

Humans Construct Their World to Suit Themselves

Human behavior is usually adaptive because humans typically build their world to suit themselves and their existing adaptations (Odling-Smee et al., 2003). As they evolve, humans continuously construct and reconstruct important components of their selective environments. In this respect, humans are no different from any other organism. Animals do not just perturb their environment at random; they build structures that are *extended phenotypes*, adaptations that allow work to be carried out more effectively (e.g., with greater yield or less energy input) and thereby incrementally increase fitness (Dawkins, 1982). Animals also deplete resources, but this often incrementally increases fitness in the short term. For instance, resource depletion is tied to life-history strategies that take account of this depletion; for instance, through dispersal or migration when resource levels are low (Odling-Smee et al., 2003). While niche construction can have both positive and negative effects on the constructor's fitness, Odling-Smee et al. (2003) are explicit about their expectation that most niche construction will be positive, increasing the short-term fitness of the constructor (although these events may well have negative consequences for other species). Even where organisms degrade their environment, it is usually because it is profitable to do so in the short term.

Moreover, as described above, animals do not just build structures; they maintain and regulate the conditions inside them, frequently to damp out variability in external environmental conditions (Odling-Smee et al., 2003). By carrying out this sort of work, organisms maintain rather than change the selection pressures upon them and, in the process, preserve the adaptiveness of their own behavior. Likewise, rather than counteracting environmental conditions, organisms may also put work into niche construction that instead changes selection pressures and initiates evolutionary episodes by perturbing components of the environment. While humans' ability to engage in this sort of niche construction is amplified by their capacity for culture, it is a phenomenon common across animal species. Like the acorn-storing squirrel or the wasp that cools her nest with droplets of water, our ancestors ensured the availability of food by tracking game and storing food and controlled temperature by manufacturing clothes and building fires and shelters. Modern refrigerator-freezers and air-conditioning are no different. Such niche construction may change environments to negate a modified or fluctuating selection pressure, thereby reducing selection.

As an illustration, imagine a population of our human ancestors exposed to an environment that becomes more arid. Our ancestors would respond by pumping or carrying in water for drinking and irrigation or relocating to a less arid region and, through this activity, would have negated selection—selection that might otherwise have generated adaptive lag. Niche construction acts to maintain environmental conditions within tolerable limits and, in the process, filters and modifies the selection acting on the *niche constructor*.

As another example, consider human artifacts. A cup is a useful drinking utensil for a human being, but it is of little utility to most other organisms lacking, as they do, the manipulative dexterity of a limb with fine motor control within easy reach of a mouth. Cups, knives, forks, spoons, jugs, saucepans, ovens, kitchen cabinets, and countless other everyday tools, implements, and artifacts are obviously specifically designed with human bodies in mind. The same holds for bicycles. Manufacturers are not constrained by their genes to produce bicycles with two pedals; rather, other designs have proved less useful to a two-footed clientele. In fact, one manufacturer did produce a one-pedaled bicycle, designed to allow Victorian women to ride sidesaddle; it was not a commercial success (Bijker, 1995).

Human culture and technology are designed to be well suited to our biological capabilities, and, as a consequence, their manufacture and utilization do not compromise our general tendency to behave in a fitness-enhancing manner. This adaptiveness is clearly not a legacy of highly specific adaptations (e.g., for milking cows or manufacturing bicycles) but rather the product of some much more general adaptations (e.g., the ability to learn, including from others, to devise solutions to problems, to teach, to communicate efficiently, to integrate information from different sources, and so forth), which collectively underlie our cultural capability. Our evolutionary legacy is not in the form of genes for specific complex processes but rather capacities for flexible knowledge-gaining and utilizing systems.

Humans Buffer Out Any Adaptive Lag through Cultural Niche Construction

For no species is niche construction more obvious and more important than our own. Our engineering and technology have tamed the planet, literally bringing light to the darkness, heat to the cold, and water to the desert. Our activities allow us to exist in a fantastically broad range of habitats. With the technology and innovations we have developed, humans are equally at home forging a living as hunter-gatherers on the baking African savanna, as fish and seal hunters in the frozen Arctic, or as white- and blue-collar workers in cities.

To a large extent, it is our capacity for culture, the transmission of information from person to person across both time and space, that makes us such potent niche constructors—the very human ability to acquire and transmit learned knowledge and skills, as well as to devise ever-more-efficient solutions to problems that build on this reservoir of shared intelligence. Other animals may possess traditions for feeding on particular foods or for singing particular songs, such as the tool-using conventions of chimpanzees or the vocal dialects of chaffinches. These animal traditions, however, lack the process of knowledge accumulation and development over generations that is the hallmark of human culture. With each technological advance, humans solve one problem but set up new challenges for descendant populations. More so than any other species, our niche construction drives, damps, and directs our evolution.

Human evolution is unique in that our culture, work, and niche construction have become self-reinforcing, with transgenerational culture modifying the environment in a manner that favors ever more culture and, with niche construction informed by cultural knowledge, becoming ever more powerful (Laland, Odling-Smee, & Feldman, 2000). A body of mathematical population genetics theory demonstrates that cultural niche construction has major evolutionary consequences and that culture amplifies the evolutionary feedback loop generated by niche construction (Laland et al., 1996, 1999, 2001; Odling-Smee et al., 2003).

Odling-Smee et al. (2003) describe two types of feedback from prior niche construction. Figures 5.1A and 5.1B illustrate the two principal routes by which a human population could respond to its own earlier activities. The first such route is via further cultural niche construction (route 1, figure 5.1A). The second route is by a genetic response to selection (route 2, figure 5.1B). Route 1 is an adaptive cultural response to a change in an environment that was brought about by earlier cultural niche construction (figure 5.1A). For example, suppose humans change their environment by polluting it. This polluted environment may stimulate the invention and spread of a new technology to cope with the contamination, alleviating the problem. Provided the response is sufficiently effective to counteract the change in the environment, the first route should be confined to the cultural level alone and should have no effect on human genetics. There would be no adaptive lag.

Consider the example of human aggregation into large sedentary communities, with the construction of towns and cities, which created, along with countless other challenges, the problem of what to do with human domestic and industrial waste products (Diamond, 1997). In such circumstances, human populations did for a short period (on an evolutionary timescale) experience novel self-induced selection pres-

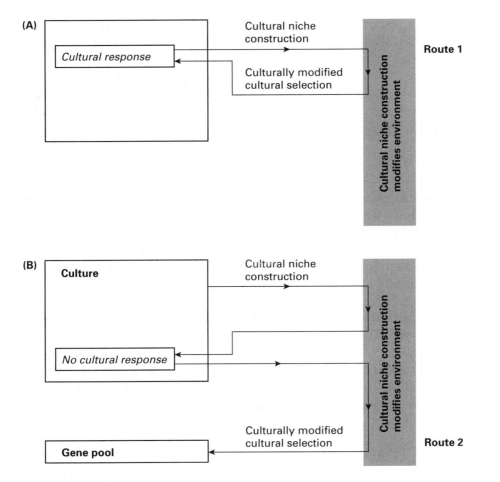

Figure 5.1
Two forms of feedback from cultural niche construction leading to (A) a cultural response or (B) a genetic response. Figure based on figure by Odling-Smee et al. (2003).

sures, different from those witnessed by their African ancestors. For instance, they were exposed to a host of diseases, including measles, smallpox, and typhoid, that thrive in dense populations with poor sanitation (Diamond, 1997). At that time, there was an adaptive lag caused by niche construction. Had these populations lacked, over time, the technology to respond to these challenges, this adaptive lag would have been maintained, the population levels would have crashed, or evolutionary change would have occurred. However, eventually, faced with disease and pollution, human populations devised solutions, including drains, sewerage plants, and water purification treatments, vaccinations, and medicines, which alleviated the problem (Diamond, 1997).

While there is nothing inevitable about the capacity of human populations to construct solutions to self-imposed problems, their capacity for culture renders human niche construction uniquely potent and fast-acting, leaving the first route by far the more likely of the two human responses to novel challenges. Other species also manifest constructed solutions to self-instigated problems; however, typically these must evolve. Such responses take place over longer periods of time and involve the populations incurring large numbers of deaths. While this, too, is a possibility for humans, the cultural solution has the advantage that it is much more immediate, with less mortality incurred.

When Humans Are Unable to Buffer Out Adaptive Lag through Further Cultural Niche Construction, Natural Selection Ensues

In recent years, biologists have been able to measure rates of response to natural selection in animals and plants. The results suggest that selection may operate faster than hitherto conceived, with significant genetic and phenotypic change sometimes observed in just a handful of generations (e.g., Dwyer, Levin, & Buttel, 1990; Grant & Grant, 1995; Reznick, Shaw, Rodd, & Shaw, 1997; Thompson, 1998). A recent meta-analysis of thousands of recorded evolved responses to natural selection concluded that the median rate of selection gradient would cause a quantitative trait to change by one standard deviation in just 25 generations, about 500 years for humans (Kingsolver et al., 2001). This suggests that significant human evolution could be measured in hundreds of years or less. Moreover, analysis of the human genome reveals hundreds of human genes showing statistical signatures of recent positive selection (Voight, Kudaravalli, Wen, & Pritchard, 2006; Wang, Kodama, Baldi, & Moyzis, 2006).

This change in perspective on evolutionary rates opens up the possibility that humans could realistically have evolved solutions to self-imposed problems over the last few millennia. This is illustrated by route 2, shown in figure 5.1B, which applies whenever human cultural processes fail to express a sufficiently effective response to an environmental change, resulting in modified natural selection pressures, which, in turn, change gene frequencies.

There are several examples of human work, particularly related to agriculture, that have impacted the frequency of genes within human populations (Odling-Smee et al., 2003; Laland & Brown, 2006). For instance, a population of Kwa-speaking yam cultivators in West Africa cut down clearings in tropical rain forests to enable them to grow their crops (Durham, 1991). The clearings increased the amount of standing water present, and such improved breeding grounds for mosquitoes increased the prevalence of malaria in the region. This increase in malaria-modified natural selection in favor of the sickle-cell S allele occurs because, in the heterozygous condition, the S allele confers some protection against malaria. The fact that other Kwaspeaker populations, whose agricultural practices are different, do not show the same increase in the frequency of the *S* allele supports the conclusion that cultural human practices can and do result in evolutionary change in specific human populations (Durham, 1991).

Once again, this evolutionary change acts to restore the adaptiveness of human behavior. Among the malaria-rife regions of the Kwa homeland, being heterozygote for the sickle-cell *S* allele is adaptive. Similarly, in a dairying society, genes expressed in high lactase activity pay fitness dividends. In this instance, niche construction has only temporarily induced an adaptive lag, this time alleviated through rapid selection, as opposed to further niche construction. Strong support for this process comes from recent analysis of the human genome (Wang et al., 2006), where the lactase persistence (LP) allele and various alleles conferring resistance to malaria (e.g., G6PD, TNFSF5) are among the genes exhibiting the strongest signatures of recent positive selection.

Implications for Human Behavioral Ecology

How do we know that the arguments presented above are correct, that the cultural innovations and technological solutions that have resulted in such widespread, rapid, and recent human niche construction are typically adaptive? One overwhelmingly compelling fact supports this assertion—human global population growth (Smil, 1993). Ultimately, if human traits are largely adaptive, that will be manifest in high fitness, as exemplified by substantive intrinsic growth rates in human populations. Conversely, if contemporary human behavior is largely maladaptive, then, globally, human numbers should dwindle in size.

Herein lies a problem for advocates of the adaptive-lag hypothesis: our ancestors did not start to thrive until after they had left their Pleistocene "environment of evolutionary adaptedness"; it is in the Holocene, after the Pleistocene, that we see the explosion in human population size and human colonization of the globe. Neither is this explosion attributable to expansion of hunter-gatherer societies, the traditional societies that evolutionary psychologists would imagine most likely to resemble ancestral societies. This population growth, from the Holocene to the present, provides the clearest indication that a large proportion of human characteristics remain adaptive even in modern constructed environments.

Even so, human behavior is at times maladaptive, with a resultant lowering of fitness. Indeed, casual observations of cultural phenomena from abstinent religious beliefs to destructive drug abuse, combined with the findings of population genetic theory, which reveals that maladaptive cultural traits can spread under a variety of circumstances (Feldman & Laland, 1996; Laland & Brown, 2002), convince us that there is nothing inevitably adaptive about human behavior. Nonetheless, based on the three arguments above, we anticipate that adaptive human behavior will be the norm and maladaptation the exception. Furthermore, we see no reason to expect greater levels of adaptive behavior in preindustrial, small-scale, or hunter-gatherer societies than in the fully industrialized urban metropolis. In spite of the massive changes humans have brought about in their worlds, the aforementioned processes collectively help to maintain a largely adaptive match between human features and the factors in their environment (Laland & Brown, 2006).

If human behavior remains largely adaptive, even in modern environments, this provides a justification, based on the fundamentals of evolutionary theory, for the widespread and general application of behavioral ecology methods to all human societies, including the most modern postindustrial societies. That is not to say that the methods of human behavioral ecology are a panacea for the study of humanity. There will probably be circumstances in which these methods are not effective; however, there is no good evolutionary reason to believe that the tools of behavioral ecology will not be successful in modern societies, in spite of the dramatic changes in our environments over the last few millennia. Furthermore, there is no evolutionary reason to expect the methods to be less successful in postindustrial than preindustrial societies.

Evolutionary Models of the Study of Work

We now consider the significance of this line of reasoning for the biological study of work. Within the fields of organizational science and industrial psychology, the study of human work focuses on the behavior of human beings in the modern workplace in postindustrial societies. As argued above, the traditional optimality methods of behavioral ecology, hitherto largely restricted to small-scale, traditional societies, could still be of utility in predicting human behavior in the contemporary workplace. The relationship between work in the thermodynamic and vocational senses is discussed below and in more detail in chapter 8.

In chapter 8 of this volume, Levin, Laland, and Saturay argue that successful lineages of organisms must engage in two classes of activity, which can be broadly categorized as net energy accruing and net energy depleting. Organisms seek to accrue from their environments (a) sufficient energy to cover the cost of the activity itself and essential homeostatic processes, and hence to survive (the *working energy* cost), and (b) an energy surplus to allocate to growth, development, mating, and reproduction (the *take-home energy* cost). From the perspective of evolutionary adaptation, successfully reproducing organisms must actually engage in such growth, development, mating, reproduction, rearing, and/or related activities, which ultimately deplete net energy reserves but nonetheless are adaptive because they are central to reproductive success.

Activities such as photosynthesis and foraging are adaptive because they typically result in net energy accrual and the production of order in biological systems. Such activities benefit organisms by recouping the energy cost of the activities themselves and of other essential energy-consuming physiological processes that are ongoing throughout the duration of the activity (e.g., respiration). These activities also generate energy stores from energy surpluses in forms ranging from complex molecules to orderly structures that can potentially allow for, or facilitate, future reproduction. Such energy surpluses bear a direct relationship to biological fitness. Turner (2000) suggests that fitness can be understood as the ability of an organism to gather and deploy energy. This energetically based fitness measure potentially provides a relevant currency for fitness that is particularly useful in modern postindustrial settings. Indeed, this energetic currency might prove to be a more accurate estimate of biological fitness than number of offspring, because it also captures the capacity of parents to invest in their offspring and grand-offspring, as well as wealth and other resources that may be passed down the generations.

Human work in the vocational sense should arguably be regarded as an example of such net energy-accruing activities. When humans go to the workplace, they do so in order to trade their exertions for tokens, such as money, that can be converted into resources, such as food, heating, or protection from the elements, and that ultimately provide them with energy in various forms. Thus, work (vocational) is that subset of work (thermodynamic) that involves activities that function to accrue an energy surplus (i.e., for which the gross energy accrued is greater than the working energy cost).

In chapter 8, Levin, Laland, and Saturay propose that the evolutionary success of organisms depends on their ability to accrue energy in a manner that minimizes the working-energy cost and maximizes the surplus that can be allocated to take-home energy under the constraints necessary to stay alive. Given a finite limit on the energy that can be accrued, greater energy expended in work (vocational) means less take-home energy. Conversely, greater energy can only be allocated to take-home energy (a) by increasing the efficiency of work, such that for a given unit of working-energy cost a greater net surplus energy is accrued, or (b) by reducing the working-energy cost. Given the inherent trade-off between these possible energy sinks, Levin, Laland, and Saturay propose that all organisms, including humans, should have evolved adaptations designed to maintain an optimal balance (or optimal range) of energy allocation to working energy and take-home energy budgets. This essential balance is called the *working energy/take-home energy trade-off*.

The significance of such a balance is that its recognition potentially sheds light on aspects of human behavior in the workplace. The working energy/take-home energy trade-off predicts that humans will act in ways that are consistent with a tendency toward seeking and maintaining a balance in energy allocation, between the conflicting demands of working and take-home energy, within an acceptable range. In chapter 8, this argument is used to develop explanations of levels of job dissatisfaction, work performance, and counterproductive behavior and leads to some nonintuitive predictions.

The working energy/take-home energy hypothesis has yet to be tested, and it remains to be seen whether it will prove important. However, we stand by the general argument: The niche-construction perspective within evolutionary biology leads to both an explicit connection between work (thermodynamic) and work (vocational), and the expectation that modern human behavior will be largely adaptive. This legitimizes the application of human behavioral ecology methods to the modern workplace and potentially provides new biological tools and currencies with which to understand human work.

Acknowledgments

We are grateful to John Odling-Smee for helpful comments on an earlier draft of this chapter.

References

Aoki, K. (1986). A stochastic model of gene-culture coevolution suggested by the "culture historical hypothesis" for the evolution of adult lactose absorption in humans. *Proceedings of the National Academy of Sciences of the United States of America, 83,* 2929–2933.

Bijker, W. (1995). *Of bicycles, bakelites, and bulbs.* Cambridge, MA: MIT Press.

Borgerhoff Mulder, M. (1990). Kipsigis women's preferences for wealthy men: Evidence for female choice in mammals? *Behavioral Ecology and Sociobiology, 27*(4), 255–264.

Cosmides, L., & Tooby, J. (1987). From evolution to behavior: Evolutionary psychology as the missing link. In J. Dupré (Ed.), *The latest on the best: Essays on evolution and optimality* (pp. 277–306). Cambridge, MA: MIT Press.

Dawkins, R. (1976). *The selfish gene.* Oxford: Oxford University Press.

Dawkins, R. (1982). *The extended phenotype: The gene as the unit of selection.* San Francisco: Freeman.

Diamond, J. M. (1997). *Guns, germs and steel: The fates of human societies.* New York: Norton.

Durham, W. H. (1991). *Coevolution: Genes, culture and human diversity.* Palo Alto: Stanford University Press.

Dwyer, G., Levin, S. A., & Buttel, L. (1990). A simulation of the population dynamics and evolution of myxomatosis. *Ecological Monographs, 60,* 423–447.

Feldman, M. W., & Cavalli-Sforza, L. L. (1989). On the theory of evolution under genetic and cultural transmission with application to the lactose absorption problem. In M. W. Feldman (Ed.), *Mathematical evolutionary theory* (pp. 145–173). Princeton, NJ: Princeton University Press.

Feldman, M. W., & Laland, K. N. (1996). Gene-culture coevolutionary theory. *Trends in Ecology & Evolution, 11*, 453–457.

Grant, P. R., & Grant, B. R. (1995). Predicting microevolutionary responses to directional selection on heritable variation. *Evolution; International Journal of Organic Evolution, 49*, 241–251.

Hansell, M. H. (1984). *Animal architecture and building behaviour.* New York: Longman.

Holden, C., & Mace, C. (1997). Phylogenetic analysis of the evolution of lactose digestion in adults. *Human Biology, 69*, 605–628.

Kingsolver, J. G., Hoekstra, H. E., Hoekstra, J. M., Berrigan, D., Vignieri, S. N., Hill, C. E., ... Beerli, P. (2001). The strength of phenotypic selection in natural populations. *American Naturalist, 157*, 245–261.

Krebs, J. R., & Davies, N. B. (1993). *An introduction to behavioural ecology* (3rd ed.). Oxford: Wiley Blackwell.

Laland, K. N., & Brown, G. R. (2002). *Sense and nonsense: Evolutionary perspectives on human behaviour.* New York: Oxford University Press.

Laland, K. N., & Brown, G. R. (2006). Niche construction, human evolution and the adaptive lag hypothesis. *Evolutionary Anthropology, 15*, 95–104.

Laland, K. N., Odling-Smee, F. J., & Feldman, M. W. (1996). On the evolutionary consequences of niche construction. *Journal of Evolutionary Biology, 9*, 293–316.

Laland, K. N., Odling-Smee, F. J., & Feldman, M. W. (1999). Evolutionary consequences of niche construction and their implications for ecology. *Proceedings of the National Academy of Sciences of the United States of America, 96*, 10242–10247.

Laland, K. N., Odling-Smee, F. J., & Feldman, M. W. (2000). Niche construction, biological evolution, and cultural change. *Behavioral and Brain Sciences, 23*, 131–175.

Laland, K. N., Odling-Smee, F. J., & Feldman, M. W. (2001). Cultural niche construction and human evolution. *Journal of Evolutionary Biology, 14*, 22–33.

Laland, K. N., Odling-Smee, F. J., & Feldman, M. W. (2004). Causing a commotion. Niche construction: Do the changes that organisms make to their habitats transform evolution and influence natural selection? *Nature, 429*, 609.

Levin, R. A., & Laland, K. N. (2003, February). The working energy/take-home energy hypothesis. In R. A. Levin, A. C. Bekoff, J. G. Rosse, and C. L. De La Rocha (Chairs), *Putting energy and information to work in living systems.* Symposium conducted at the annual meeting of the American Association for the Advancement of Science, Denver, CO.

Lewontin, R. C. (1982). Organism and environment. In H. C. Plotkin (Ed.), *Learning, development and culture* (pp. 151–170). New York: Wiley.

Lewontin, R. C. (1983). Gene, organism, and environment. In D. S. Bendall (Ed.), *Evolution from molecules to men* (pp. 273–285). Cambridge: Cambridge University Press.

Naiman, R. J., Johnston, C. A., & Kelley, J. C. (1988). Alterations of North American streams by beaver. *Bioscience, 38*, 753–762.

Odling-Smee, F. J. (1988). Niche constructing phenotypes. In H.C. Plotkin (Ed.), *The role of behavior in evolution, 73–132.* Cambridge, MA: MIT Press.

Odling-Smee, F. J., Laland, K. N., & Feldman, M. W. (1996). Niche construction. *American Naturalist, 147*, 641–648.

Odling-Smee, F. J., Laland, K. N., & Feldman, M. W. (2003). *Niche construction: The neglected process in evolution. monographs in population biology. 37.* Princeton: Princeton University Press.

Reznick, D. N., Shaw, F. H., Rodd, H., & Shaw, R. G. (1997). Evaluation of the rate of evolution in natural populations of guppies (*Poecilia reticulata*). *Science, 275*, 1934–1936.

Richerson, P. J., & Boyd, R. (2005). *Not by genes alone: How culture transformed human evolution.* Chicago: University of Chicago Press.

Smil, V. (1993). *Global ecology: Environmental change and social flexibility.* New York: Routledge.

Smith, E. A., Borgerhoff Mulder, M., & Hill, K. U. (2001). Controversies in the evolutionary social sciences: A guide for the perplexed. *Trends in Ecology & Evolution, 16*(3), 128–135.

Symons, D. (1987). If we're all Darwinians, what's the fuss about? In C. Crawford, M. Smith, & D. Krebs (Eds.), *Sociobiology and psychology: Ideas, issues and applications* (pp. 121–146). Hillsdale, NJ: Erlbaum.

Tebbich, S., Taborsky, M., Febl, B., & Blomqvist, D. (2001). Do woodpecker finches acquire tool-use by social learning? *Proceedings. Biological Sciences, 268*, 2189–2193.

Thompson, J. N. (1998). Rapid evolution as an ecological process. *Trends in Ecology & Evolution, 13*, 329–332.

Tooby, J., & Cosmides, L. (1990). The past explains the present: Emotional adaptations and the structure of ancestral environments. *Ethology and Sociobiology, 11*, 375–424.

Turner, J. S. (2000). *The extended organism: The physiology of animal-built structures.* Cambridge, MA: Harvard University Press.

Ulijaszek, S. J., & Strickland, S. S. (1993). *Nutritional anthropology: Prospects and perspectives.* London: Smith-Gordon.

Voight, B. F., Kudaravalli, S., Wen, X., & Pritchard, J. K. (2006). A map of recent positive selection in the human genome. *Public Library of Science Biology, 4*, 446–458.

Wang, E. T., Kodama, G., Baldi, P., & Moyzis, R. K. (2006). Global landscape of recent inferred Darwinian selection for *Homo sapiens*. *Proceedings of the National Academy of Sciences of the United States of America, 103*, 135–140.

Whiten, A., Goodall, J., McGrew, W. C., Nishida, T., Reynolds, V., Sugiyama, Y., … Boesch, C. (1999). Cultures in chimpanzees. *Nature, 399*, 682–685.

Williams, G. C. (1966). *Adaptation and natural selection: A critique of some current evolutionary thought*. Princeton, NJ: Princeton University Press.

Williams, G. C. (1992). Gaia, nature worship, and biocentric fallacies. *Quarterly Review of Biology, 67*, 479–486.

What kind of human work would provide an example of the ways in which humans incorporate both variability and creativity into work? The work of design is one such example. Design researcher Alan Blackwell provides a unique exploration of the work of design by describing an interdisciplinary study in which designers from disparate fields were studied by (and studied with) a group of researchers, also from disparate fields, each of whom was also a designer. This rich setting yields many counterintuitive insights on how design is work—which aspects of design seem unique to the domain, which seem to bring out the quintessentially human aspects of design, and which look much the same as any day-to-day work. Blackwell concludes with a kind of design that each of us engage in every day, attention investment, *the conscious and unconscious work by which we decide how to shape and move through our environment.*

6 The Work of Designers: Cultures of Making and Representation

Alan Blackwell

This chapter is concerned with a particular kind of contemporary human work: the professional work done by designers. This choice might at first seem an overly specialized area of work for exploration. However, design is a useful focus in several respects. In this chapter, along with exploring the work of design as an interesting area in itself, we offer the possibility of treating design work as a "model system" for studying professional human work, just as a fly's eye (chapter 2) and a chip foundry (chapter 3) have served as model systems in exploring different aspects of work in living systems.

This chapter builds on two areas of comparative study. First, we report results from a research project that set out to compare the range of professional design work by drawing on descriptions of design work provided by professional designers themselves. The results are presented as a number of themes exploring the general nature of design. Second, we pay special attention to the ways in which design work is evolving and developing through the use of computers; these changes also offer parallels to other processes of change in living systems.

Design work is not simple physical rearrangement of the world; it includes processes of planning and imagining the results of physical activity. Design work is not purely intellectual, either—even the planning stages of design often involve the construction of models, plans, or other representations of the final product and its parts.

The work of designing physical artifacts and tools involves a process of translating a conceptual requirement via the mental image of a desired form into an actual object. As an example of this inherent intricacy and interplay in the work of design, consider a sample design project that is described in more detail later in this chapter. In this project, a car designer started his work by first persuading a client that a group of customers was not served by any of the models in the company's existing range. Next, he proposed a new market segment, and persuaded the client literally to buy into a concept of a car that would appeal to that market. After that, he worked with a sculptor to define the form of that car, researched manufacturing techniques, and invented novel uses of materials and fastenings. Finally, he supervised the construction of prototypes, analyzed their mechanical performance, refined the manufacturing process, and orchestrated the press launch. As a successful designer, he worked in all of these widely varying domains, just in the scope of one project. It is for this reason in particular that understanding the work of design can be an effective system for helping to build an understanding of work.

The Changing Work of Design: An Overview

Design, in the broadest sense (e.g., Heskett, 2002), is the intellectual process by which humans modify the material world for our own benefit. All organisms modify the world through niche construction, as described in chapter 5. However, as noted above, design involves thinking about the intended design outcome. In order to do so, it is necessary for human designers either to imagine the thing being designed to create an internal mental representation or to make an external representation, such as a plan or model. In preindustrial technological eras, design plans or models were turned into actual products by a person with craft skills, often the same person who had made the plans.

In contemporary manufacturing economies, however, the work of design has increasingly become remote from craft skills for shaping material. When the materials are shaped by machines, designers' work instead involves specifying machine behavior, often using artificial design languages to program a computer-operated machine. These are still design representations, but they take the form of abstract specifications of machine behavior, rather than representational pictures or physical models (Blackwell, Hewson, and Green, 2003). Even in design professions in which pictures are still necessary, these are often prepared on computer screens, with the result that general characteristics of computer representation have distinctive implications for the future of design, as described in the final section of this chapter.

The distinctiveness of individual design professions, once determined by an individual's skill in the use of specific tools and materials (echoing early specialization of skilled crafts work), is being replaced by a working environment that tends to be surprisingly similar across different domains of design, within which design professions are separated more by conventions of social status and education than by their repertoire of manual skills. Such trends in professionalization and the social and economic contexts of design have further transformed the modern work of design as much as have innovations in modeling and representation.

A Comparative Approach to the Study of Design

In the Across Design Project, a joint venture between researchers at Cambridge University and the Massachusetts Institute of Technology (MIT), we set out to discover which things are similar and which are different across the range of design work done by different people (Blackwell, Eckert, Bucciarelli & Earl, 2009). In fact, not all of the participants would describe their own work as "design." However, we wanted to cover as wide a range as possible of professional design work, so we focused on recruiting a diverse sample of expert participants, regardless of how they themselves might describe their professional involvement with design. The 24 participants included fashion designers (a pattern designer and a couturier), architects (designing public housing, private housing, and public assembly spaces), engine designers (jet engines and diesel engines), automotive designers (high-volume trucks and low-volume sports cars), multimedia designers (university courses and commercial Web sites), software designers (large government systems and single-user programming languages), and designers of healthcare products (new drug compounds and injection devices for administering drugs), as well as an industrial product designer, a civil engineer, a filmmaker, a graphic designer, a food product designer, a packaging designer, an electronic product designer, and a furniture designer.

The team of researchers brought its own diverse range of perspectives. We came from academic affiliations in engineering, architecture, computer science, and other fields. In addition, all of us were ourselves designers or teachers of design in some domain. We shared a desire to develop a broader description of design work than is conventionally applied in the field of design research, and this desire shaped our approach to the project.

We began our work with the designers by providing them with a diagram that showed the categories of our own interests in learning about design work but warned them not to treat this as proscriptive or as a questionnaire. We instead asked designers to present a case study of their own work, describing a single design project. We asked only that their choice of case study illustrate what they believed the most salient aspects of their own profession. Each presentation was followed by discussion among the designers, so that other designers might freely draw comparisons to their own experience, for their own benefit and for ours.

The Contemporary Work of Design

Based on the discussions and analysis of the Across Design Project discussions, we identified several broad patterns in the contemporary work of design (Eckert, Blackwell, Bucciarelli, & Earl, forthcoming). The predominant theme, one that surprised even those of us with extensive contact with the design professions, was the extent to which the design and construction of material artifacts is a social process. The designers we spoke to spent little of their time working with their hands or even analyzing the form and function of their products. Instead, the bulk of their work involved social negotiation with other parties: users, clients, public bodies, and regulatory authorities.

The next sections of the chapter describe the common patterns that we found. We present these in an order that moves from the most fundamental internal thought processes of an individual through increasingly complex and far-reaching social and organizational processes. While some aspects of the experience of individual designers have been well established for centuries, these social and organizational patterns of work have been more commonly observed in the past few decades. The six patterns of work we found in the work of design are:

1. the creative step of identifying an opportunity for innovation,
2. the step of converting a concept into material form through the use of models,
3. the need to coordinate with the work of others when constructing more complex products,
4. the work of negotiation that is introduced when social specialization separates the roles of designer from those of product users and client,
5. the context of regulation and economic forms and their consequences for the work of design, and
6. the work by which the design professions themselves are maintained as social structures.

1. Creativity

The first step in any design process is the conception of a need for something new. Much routine manufacture in modern economies could proceed without conventional design work. Attractive and functional patterns for most common products were established decades or even centuries ago. The clothes, chairs, tables, cups and spoons, papers, and pencils of everyday life have remained much the same for 50 years, without any significant difference in their suitability for end user needs or in the real prices paid for them. Now that manufacture allows the mundane duplication of established products like these, the work of designers is focused on novelty or improvement—innovation. In the industrial age, the most admired design work has evolved from the display of craft skill to the display of creativity—the pursuit of innovation.

The novelties that contemporary designers pursue include both pragmatic improvements, such as reduced cost, greater efficiency, or new functionality, and also the demands of fashion or styling that are constituents of the work of culture. The former improvements are empirically observable and somewhat predictable in an engineering fashion. The latter improvements depend on changing popular taste and include an element of surprise (Coates, 2002). Students entering traditional design professions, such as clothing, graphics, or furnishing design, are attracted to style and fashion as central components of design work. They are taught to collect aesthetic trends in a systematic way—for example, using *mood boards*, visual scrapbooks that define contemporary or desirable style cues. The packaging designer participating in the Across Design Project talked about the use of mood boards in the design of detergent bottles. In this market, the same products are repackaged periodically to retain market interest. His company therefore employs an advertising agency to construct a mood board that will reflect the target market and inspire his design. These contemporary trends or moods also spill across product and industry sectors, as noted by another Across Design Project participant who works in graphic design. In the project she described, a brochure for the youth market applied a fashion trend that was popular with her own young students at the time—using images taken originally from photographs with television scan lines superimposed so that they appeared to have been captured from a video feed. The result has no functional benefit (the image quality is worse), but it offers an association with contemporary media.

As compared with innovation for novelty, functional innovations—the proverbial "better mousetrap"—are apparently much more difficult to produce from the kind of data typically provided to designers by such methods as styling cues, mood boards, and market research. One industrial product designer told us that, instead, his creative insights are achieved from "three gin-and-tonics and a hot bath." However, this kind of appeal to tacit knowledge and to mystery in the creative process may be disingenuous. In particular, ownership of a mysterious process that cannot be duplicated serves the needs of the professional design consultant by ensuring a stream of future work. Indeed, the same participant's definition of a successful project was one after which a happy client returns to him with further work. The critical component of such success, in his experience, is the achievement of a creative "X-factor" that cannot exactly be described but that leads the customer to have a "love affair with the product." Just as with detergent bottles, this designer must work with marketing departments to imagine that very customer-in-love, taking a creative leap into the unknown stylistic future—often dragging his clients in the manufacturing departments reluctantly behind him. Central components of this work are empathy and imagination, shared with both customers and clients.

2. Models and Representations

After the creative step of conceiving a new product idea as an internal mental representation, designers immediately begin to construct external representations, such as plans and models. These external representations are, of course, necessary for the abstract internal conception of the designer to be communicated to clients, users, and manufacturers. More surprisingly, the designer will often use external representations in conversation with him or herself in order to explore and refine the design concept (Schon, 1983).

As one example of creating external representations, the act of sketching is an important component of the individual creative process. A typical cycle of exploratory design might start with the designer making a primitive sketch that is little more than some rough, ambiguous marks with a pencil. On looking at those pencil strokes after they are set down, the designer then often sees them in a different way, reinterpreting them to provide further ideas or inspiration. Those, in turn, might result in a further sketch, which can then be reinterpreted again, exploring configurations until the best solution is found. A designer of housing developments told us, for example, that he made hundreds of drawings he called "massing sketches" in order to explore the ways that the three-dimensional volumes of the individual houses would combine into larger scale physical structures.

This common strategy of using sketches as a creative tool for design work is a trick of *distributed cognition*—the situations in which reasoning processes are accomplished not only in the head but through the combined resources of the brain with representational marks (writing or drawing). Common examples of distributed cognition in other kinds of human work include shopping lists (distributed memory), train timetables (distributed planning), and mathematical pencil workings (distributed calculation). The aim of the sketching trick is rather more subtle—it is widely considered that sketching allows the designer to escape any lexical preconceptions that might constrain his or her imagination with established concepts. Instead, through sketching, the designer can escape into a world of the visual—a world where marks contain surprises (Eckert, Blackwell, Stacey, & Earl, 2004).

Sketching is much admired as a way of solving creative problems, not only among those designers whose work is associated in the popular imagination with artistic use of a pencil (e.g., graphics, clothing, architecture) but also by our project participant who designed jet engines. He told us that he encourages his engineering staff to work quickly with a pencil and develop a sketch, or several, in order to address the challenge in his industry of turning a quantitatively described analytic problem statement into a mechanical solution expressed in the physical form of an engine.

Even in industries where paper and pencil sketches are uncommon, other means can be used to explore and iterate ideas in a form that is external to the designer's own thoughts, with three-dimensional sketches providing an opportunity for conversation with materials. For example, a furniture designer told us how she went about the process of creating a lampshade that is now a popular product in the United Kingdom, made of folded and formed plastic. She developed the shape by working like a sculptor, exploring sheets of the material with her hands until she found a form that was both attractive and practical. In a very different approach, one of our car designers described the essential sculptural work of a key member of his design team, a specialist clay modeler, whose role in the design project was specifically to create a one-eighth scale clay concept model of the car that they could view, discuss, and refine. However, this model was not purely exploratory: It also became a sales tool, intended to sell the design concept to a client, with the aim of getting funding for the more routine and detailed work of mechanical and production design.

In each of these examples, we learned that sketches, models, and other representations are not purely individual tools for doing design work but are communicative tools as well. The clay model of the car was an essential prop for the communication of the designers with clients, and it also provided a communicative tool among the members of the design team. The central collaborative relationship on the team was between the car designer and the model maker, with their communication taking place around the model, as it took shape between them. The subtleties of visual and physical form for a design are not necessarily expressible in words, so representational tools, such as the development of the clay model, form a language and a way of doing work among members of a design community.

When designers present their ideas to clients or to the public, even designers who already make full use of computer facilities may return to traditional paper drawing. An architect who regularly presented preliminary design work to members of the public told us that he would take computer renderings of the design and then trace over them with colored pens before a public meeting to produce handcrafted representations. In part, he did this to add atmospheric effects unavailable on the computer, for example to help future users imagine what the development would look like in different seasons of the year or times of day. But he also wished to make a more subtle point: Rather than a dehumanized computer rendering, the pen drawing is a human artifact, open for discussion (Bresciani, Blackwell, & Eppler, 2008). Furthermore, a colored perspective drawing is clearly an artifact of skill, establishing his status as a craftsman deserving respect for that skill when he meets the public. These negotiated aspects of design work extend the role of sketches beyond personal creativity, to the construction of shared modes of understanding and even new graphical "languages."

3. Collaboration over Periods of Time, and Managerial Processes

Almost all of the design projects that were described to us involved collaboration between designers and teams of technical specialists. These collaborations extended over periods of months or years. The uncertainties inherent in creating a novel product mean that any aspect of the process may take longer than expected. Uncertainty also means that connections between different components or the relative sizes and properties of components may change over time. When different components are subcontracted to different teams or companies, such changes can have domino effects elsewhere in the design process. Under such circumstances, a great deal of design work is, in fact, the work of project management. The most common shared representations described to us during the project, other than representations of the product itself, were managerial coordination tools such as Gantt charts (a project timeline broken down by activity and goals), showing coordination between different aspects of the work to be done over the lifetime of a project.

Technical factors, including changes to design requirements (due to market changes or clients changing their minds), are the most significant threat to project timescales, especially as the result of domino effects propagating changes to other components. In industries where the functionality of products is expected to change, especially in the software industry, many design management strategies are intended to minimize the technical consequences of functional change. A developer of large software systems described the way that multiple versions of the product are delivered, with each cycle of refinement short enough that any necessary change will be discovered early, so that managers can plan around them and not compromise final deadlines. But requirements can change in all design professions. An architect who specialized in community-managed projects, such as churches and schools, had to allow for the fact that her clients were often inexperienced with building projects and might not be aware of the importance of maintaining an agreed-upon design brief. She therefore took care to educate them regarding the stages of the design process, and had them sign off on design phases so that committees of client representatives (who were often volunteers) would recognize and acknowledge the prior decisions they had made.

Within larger organizations, agreements on design parameters or on the interfaces between components are often made between several departments of a company. The achievement of a complete design model must be recognized by formal agreements freezing specified properties for official transmission between two departments or transition between two development phases. In the car industry, for instance, these include the transition from advanced development to product design phases that recognize the decision to invest funds in the more costly final design (in the case of large companies, carried out in a different specialist department). The achievement of an agreed-upon model is often described as a gateway point that must be approved both by design teams and by senior managers so that the consequences of further unexpected change can be made accountable. In industries with extremely complex and expensive design processes, such as jet engine design, these gateway deliveries of partial designs indicate the range of allowable variability in important design parameters (chapter 3). These parameters are known as *cardinal data* and are often carefully managed to reduce the risk of unexpected change; for example, by a team agreeing to supply an estimated value immediately, along with an agreed-upon date on which the final value will be fixed.

The consequences of unexpected specification changes in these highly technological industries are likewise managed in a technical way. One designer reported the use of Monte Carlo simulations to anticipate possible specification changes and their consequences, while another described a formal risk quantification process in which every possible variation is assigned a relative likelihood from 1 to 3, which is then multiplied by the severity of the consequences should it occur. This type of engineering work is a process of making human relationships and conversations systematic through formal description and systems that utilize those formal descriptions as a way of developing agreed-upon actions and objectives that will be embodied in the final product.

4. Social Networks: Clients and Users, and Negotiation

Much of modern human work depends on the social context in which it is achieved. As described in chapter 7 of this book, getting the work done is inextricably linked to motivations and obligations grounded in the worker's social environment in the workplace and elsewhere. In contrast to the extensive interaction within the design workplace, however, the professional designers we met in the Across Design Project have surprisingly little contact with the end users who provide the business motivation for the products they design. In this respect, their work is unlike that of traditional craft professions, where a craftsman/designer is commissioned to create a piece of work that will meet the needs of a particular client.

A more typical economic chain in contemporary design begins when an independent designer is awarded a commission to work on behalf of a manufacturer. The *design brief* (a term meaning instructions from the client, as with a legal brief) that is provided to the designer might be based on market research conducted by the client, including surveys of the eventual users or customers, but it was unusual for our participants to ever meet the end users themselves. Instead, fashions or design trends, represented in the marketing surveys, appeared to act as surrogates of user needs. Perhaps the most vivid example of the separation between market-influenced design and the reality experienced by end users was provided by our car designer. In the 1970s, the designer had persuaded his client that the British car market of the time had an unrecognized demand for a "sporty three-wheeler," because three-wheeled cars were then subject to less stringent driver's license regulations. Unfortunately, the innovative layout brought some unexpected dynamic characteristics once the model was produced: At the press launch, motoring journalists rolled several cars! The Bond Bug went on to be a cult classic, with a loyal fan club, but it was not a great commercial success.

Like design collaborations, designer–client relationships also change over time. If a designer works for the same client over many product cycles, whether as an independent designer or as an employee of the manufacturer, then a closer relationship between design and market than just described may become possible. This is particularly valuable in industries in which products are repeatedly revised over periods of many years, as in the design of mass-market automobiles or complex commodity components. Our diesel engine designer described a highly developed design structure of this kind. His market data included feedback on product reliability and lifetime operating costs, allowing his team to make incremental improvements that would benefit future users. However, the end users of his products are still at one remove from the designer: Diesel engines are sold to vehicle manufacturers, who gather performance and maintenance data from their customers. This statistical data, rather than contact with end users, drives this design process.

Our second engine designer, responsible for passenger jet engines, described an analogous process. He told us that he regards statistical and financial data as a component of the product. When airlines lease a plane, they specify engines that match their expected passenger profile, including journey time and distance, fuel consumption, and periods of maintenance. Lease arrangements for passenger jets specify the total cost of operation, including the actual costs of labor and parts over the whole life of each engine, rather than using statistically based estimates of reliability and service costs. Thus, a design or manufacture that is at variance with the initial specifications can be financially ruinous to the leasing aircraft owner, who ends up bearing the costs that the airline has not paid for in the lease. To compete in a market with this kind of complexity and risk, the engine designer must work with a team of financial analysts, insurance specialists, and maintenance contractors to define the entire product as a carefully specified package that goes far beyond the already complex requirements for an engine to lift a plane into the air.

The working relationship between designer and design client can actually become even more complex in fields that combine large-scale financial or legal processes, such as those just described, with substantial demand for large-scale creative innovation. The work of architects competing for large-scale corporate and public projects exemplifies a design domain of this kind. Architectural practice often addresses these challenges by segregating these conflicting factors into separate project phases. A project often starts with a design competition among architecture firms, whose winner must form a relationship with a construction company after the competition phase for detailed planning. After the competition and the award of commission, the position of the two companies is then reversed: The building contract was originally made between the client and the architect, with the construction company appearing as a subcontractor to the architect. After a process called *novation*, the contract is converted into one between the client and construction company, with the architect becoming a subcontractor. The original design competition is a contest of creativity. But once construction starts, creativity is no longer the top priority. This change results in intricate legal changes of liability, financial accountability, and contractual authority to make modifications to the design.

The competing design priorities contributing to this idiosyncratic arrangement also include social factors around civic pride and corporate leadership. There is a tension between civic ambition to exhibit wealth and power and at the same time being responsible stewards of the public or corporate funds. The architect and the architect's design must tread a fine line between practicality, opulence, and parsimony. Where it is necessary that design products be both creative and practical, design work must be organized in a way that allows creative and practical kinds of people to work together comfortably, even where their priorities and work practices might be directly opposed to each other. The process of novation is one example of a legal convenience that allows these different approaches to the work of design to be combined in the public interest.

5. Society, Markets, and Regulation

Many designed products have impacts on society that extend far beyond the business of the designer's client and the experience of the end user. To the extent that design work represents human transformation of the world, the work of design is central to social enterprise: We found many situations in which the designer's work was to interpret public policy in the material world. Designers in the transport industries (aviation, automobile, and freight) are highly constrained by environmental regulations on noise and emissions, for example. Public policies on exhaust emissions such as greenhouse gases often set targets in many countries over 5 or 10 years. Engine designers told us that their work has become dominated by environmental legislation. This means that a particular client's requirements of cost, performance, and functionality can only be accommodated to the extent possible after regulatory targets have been achieved.

Forced design innovation in response to environmental policy can arise in more unexpected quarters. Our food designer had been commissioned to replace a traditional ice cream, whose "mouth feel"—an important dimension of customer satisfaction with ice cream—resulted from the use of a rare alpine flower. A declining species population in the wild led to the flower being designated as an endangered species, with the result that production of the ice cream became illegal. Local artisans continued covert manufacture, but price and availability were prohibitive. Our designer was commissioned to study the traditional production process, analyze its chemical structure, and develop a synthetic alternative. The result, rather than being an unwelcome industrial intervention, successfully preserved both a valued element of local culture and a broader environmental agenda.

In addition to legislative constraints on the production of goods for an otherwise free market, state intervention can provide valuable opportunities for designers. In the case of public housing schemes, the designer can be an advocate of product quality, on behalf of end users who are intended to be beneficiaries of state intervention but may be only indirectly represented within formal review processes. All large housing developments in the United Kingdom are required by law to provide a certain proportion of affordable dwellings, which may be managed by a housing association established alongside privately owned housing. Developers might be tempted to construct these affordable homes to a minimum cost standard, but one of the architects in our project described his pride in maintaining high-quality standards throughout, both as an advocate of the lower income residents and also an eventual recipient of design awards from architectural peers. This illustrates the way in which good design work is defined by a professional community, as discussed in the next section.

6. The Work of Professional Communities

Our participants were deeply concerned with the future continuity of their professional communities, both as advocates of progressive change and guardians of skilled tradition. Their work included educating young designers and lobbying public policy or professional organizations in the interests of their peers. This emphasis was particularly apparent in fields where international competition was devaluing traditional design values or technological change resulted in the loss of traditional skills.

We spoke to one of the world's leading pattern designers—the process in which a garment is broken down into individual fabric shapes to be sewn together. After 70 years of professional life, she gave us virtuosic demonstrations of perfectly tailored dresses in unusual fabrics. However, her main professional concern now was that cheap global clothing manufacture was replacing such traditional skills—the "conversation with materials" so central to creative design. Like several of our more experienced participants, she was devoting the remainder of her working life to education, preserving the knowledge accumulated in her own career.

We found that professional design communities thereby shape and preserve not only the methods used in design work but also the desired outcomes of that work. In fact, we were surprised at how many designers chose their standards of evaluation from professional peers in preference to evaluation of their work by users, markets, or clients and beyond that necessary to secure future contracts. The public housing architect described above, although taking personal pride from seeing people living happily in his projects, defined his real criterion for success as recognition from the professional (public-funded) NGO, the Commission for Architecture in the Built Environment.

Perhaps the most telling example of a designer whose audience was not the end user was the furniture designer who, asked to tell us about her most successful work, chose to tell us not about a chair that one could, in fact, sit in. Instead, she described a creation of hers developed solely for the enjoyment of other designers and design connoisseurs at the Milan Furniture Fair: this was a series of "chairs" that were actually witty sculptures—mutations of classic chair forms—most of which were impossible to sit on. Her design work, although taking physical form as exhibition pieces, did not have a physical function—it was a commentary on design work to be appreciated by other designers.

Although the furniture designer's professional methods and the outcomes of her design work are material ones, her reflection on her own experience of work drew attention to those materials and skills required for representation and communication. In the final section of this chapter, we return to this question of representational skill in the context of future design work.

Abstract Design Work and Representation Use

What kind of work is design, and how does it relate to the other kinds of work described in this book? At the start of this chapter, the work of design was described as a model system for understanding the mix of intellectual and physical work that distinguishes human work from other physical and biological work. Design is a pervasive enterprise within the material culture of human existence—our clothes, buildings, food, vehicles, and utensils. Yet contemporary design work is no longer itself primarily a material enterprise. For many designers of manufactured products, the only materials that they handle are the tools of creating representations: pencils, paper, and models. All of these, in turn, are chosen to minimize the physical work that would otherwise be required if the final products had to be built and rebuilt through trial and error, or if physical modifications were constantly required in response to changes in requirements. That is the power of representations: to explore alternative ideas in ways that are not constrained by the costs of physical action. The physical energy of the designer is expended more than anything on social activity that uses representations, such as negotiation, education, and coordination.

The Work of Representation Use

Just as the work of design can be regarded as a particularly sophisticated form of niche construction (chapter 5) that is mediated by representations, so the work of representation use can itself be compared to the other themes of this book. Representation use might involve less physical work than building a material product. The intellectual activity involved in representation use and social processes still demands the allocation of scarce resources: Time spent concentrating on a representation or attending to a conversation uses energetically expensive and valuable neural energy (chapter 2).

Fortunately, when representations are contained within computers, it is often possible to optimize cognitive resource allocation through automation. Many sophisticated computer-aided design (CAD) tools and project management systems include special-purpose programming languages that allow repetitive actions to be automated; for example, checking design calculations, propagating the domino effects of requirements changes, or converting sketches into estimated cardinal values for transmission to other design teams.

But automation does not come free. When designers become programmers through the use of powerful CAD tools, they must use yet another representation—the programming language—which is even more abstract and further removed from their material product than the CAD system itself was. Writing computer programs, even routine ones, is a design activity too, and one that requires cognitive resources in itself. In some cases, specifying an abstract automated alternative to a repetitive operation might require more concentration than simply carrying out the repetitive task (Blackwell, 2002). Furthermore, the abstract alternative could have an error, such as a program bug, resulting from the designer not concentrating hard enough. Bugs might damage the original design representation instead of improving it, requiring further concentration to repair the effects.

The use of digital tools to transform representations through programming therefore requires an investment decision about time and attention, weighing relative costs and likely risks. This decision also offers "profits," in terms of time that might be saved from attending to mundane or repetitive representational tasks and possible "losses" of attentional resources through the intellectual effort of constructing an abstract specification. This *attention investment model* of abstraction use (Blackwell, 2002; Blackwell, Rode, & Toye, 2009) is characteristic of digital tools and continues to transform professional work. Many contemporary professionals, whether designing material products or "designing" financial instruments, find themselves in the position of end-user programmers who not only create representations as a part of their work but must manipulate those representations in increasingly abstract ways (McCullough, 1998). Where work is mediated by representations, as much professional work is today, redesigning these representations has the consequence of redesigning the work, through automation and reconceptualization.

Attention investment is, in addition, one more means by which we can modify the balance of "working energy/take-home energy" (chapter 8)—this time changing the balance of energy required for intellectual work, both intellectual work in the present moment and, through computer programming, intellectual work that might be required in the future. Representation users make choices and trade-offs between doing representation work now, with immediate results (described as *direct manipulation* in human–computer interaction research (Shneiderman, 1983)) or investing in future automation and efficiency with more abstract representations by doing programming.

Beyond Design: The Future of Design Work in the Digital Era

This chapter has explored ways in which the guiding processes of building artificial environments and products for humans, of human niche construction in the industrial era, have been developed through design work. Building on that foundation, we have also seen how the work of design is conducted by individual designers and how, from designer to designer, that work is, in large part, a matter of the work of representation use.

When design representations, such as sketches, models, project plans, interface specifications, and construction contracts, are manipulated with computers, the representations themselves can be redesigned in ways that change the work of design (Blackwell & Green, 2003). This fundamental shift has placed design work at the center of the digital revolution, as the power of the use of digital representations increasingly changes many parts of human life (e.g., digital film production, business management using spreadsheets, socializing in virtual online worlds). The shift from concrete to abstract that we have seen in the work of design results in attention investment trade-off decisions, because it becomes necessary for the designer to anticipate the future interpretation of abstract notations rather than simply observing what he or she has already made. This shift is likely to become increasingly ubiquitous in all human work involving computers and digital representations. The shift away from material work and representations toward increasingly abstract and mathematical work practices is nevertheless still rooted in fundamental work processes, from the production of the silicon chips that process the abstract representations (chapter 3) to the valuable neural energy (chapter 2) that technology users must allocate when making their attention investment decisions.

Acknowledgments

The author is grateful to the many collaborators cited here and to the participants in the Across Design Project. That project was funded by the Cambridge-MIT Institute. Several studies of abstract representation and attention investment were funded by the Engineering and Physical Sciences Research Council.

References

Blackwell, A. F. (2002). First steps in programming: A rationale for Attention Investment models. *Proceedings of the IEEE Symposia on Human-Centric Computing Languages and Environments*, 2–10.

Blackwell, A. F., & Green, T. R. G. (2003). Notational systems—The Cognitive Dimensions of Notations framework. In J. M. Carroll (Ed.), *HCI models, theories and frameworks: Toward a multi-disciplinary science* (pp. 103–134). San Francisco: Morgan Kaufmann.

Blackwell, A. F., Hewson, R. L., & Green, T. R. G. (2003). Product design to support user abstractions. In E. Hollnagel (Ed.), *Handbook of cognitive task design* (pp. 525–545). Mahwah, NJ: Lawrence Erlbaum Associates.

Blackwell, A. F., Eckert, C. M., Bucciarelli, L. L., & Earl, C. F. (2009). Witnesses to design: A phenomenology of comparative design. *Design Issues, 25*(1), 36–47.

Blackwell, A. F., Rode, J. A., & Toye, E. F. (2009). How do we program the home? Gender, attention investment, and the psychology of programming at home. *International Journal of Human-Computer Studies, 67*, 324–341.

Bresciani, S., Blackwell, A. F., & Eppler, M. (2008). A collaborative dimensions framework: Understanding the mediating role of conceptual visualizations in collaborative knowledge work. *Proc. 41st Hawaii International Conference on System Sciences (HICCS 08)*, 180–189.

Coates, D. (2002). *Watches tell more than time: Product design, information, and the quest for elegance.* New York: McGraw-Hill.

Eckert, C. M., Blackwell, A. F., Stacey, M. K., & Earl, C. F. (2004). Sketching across design domains. *Proceedings of the Third International Conference on Visual and Spatial Reasoning in Design (VR'04).*

Eckert, C.M., Blackwell, A.F., Bucciarelli, L.L., and Earl, C.F. (forthcoming). Shared conversations across design. *Design Issues.*

Heskett, J. (2002). *Toothpicks and logos: Design in everyday life.* Oxford: Oxford University Press.

McCullough, M. (1998). *Abstracting craft: The practiced digital hand.* Cambridge, MA: The MIT Press.

Schon, D. A. (1983). *The reflective practitioner: How professionals think in action.* New York: Basic Books.

Shneiderman, B. (1983). Direct manipulation: A step beyond programming languages. *IEEE Computer, 16*(8), 57–69.

Joseph Rosse and Stacy Saturay interweave research on a variety of issues related to job satisfaction and dissatisfaction and their potential effects on such areas as behavior, health, and work–life and work–family conflict. Combining the perspectives of research in industrial and organizational psychology and in management with other disciplines, they begin with ways that human health has been shown to be affected by stress at work in modern organizations. They develop two areas in depth that show the effect of the workplace on human well-being. The first is adaptive behavior *research, showing that workers actively seek to psychologically adapt to dissatisfying situations at work; for instance, by active problem solving. One implication of this research is that behavior in the workplace that is commonly labeled as counterproductive is, in fact, a highly adaptive and predictable response when situations producing workplace dissatisfaction cannot be improved. Rosse and Saturay then extend their approach to a redevelopment of the challenges often referred to as* work–life balance *or* work–family conflict, *which can themselves be a continual source of dissatisfaction—and perhaps also of adaptive behavior of different kinds.*

7   Working on the Edge Today: Dissatisfaction, Adaptation, and Performance

Joseph Rosse and Stacy Saturay

A newspaper article described an interesting trend in the Denver, Colorado, metro area. Despite a 12% increase in the Denver population during the prior seven years, there was a 31% decrease in traffic citations and a 50% decrease in arrests for driving while under the influence of alcohol (Kilzer, 2005). The reason had nothing to do with Denver drivers becoming more law-abiding. Rather, the cause was attributed to a morale problem among police officers, who complained of being overworked and unsupported by senior police officials and who responded with cutbacks in their efforts.

An article in the Archives of Internal Medicine *reported that lower ranking British civil servants had almost three times the likelihood of developing Type II diabetes—a so-called lifestyle disease—than their higher ranking colleagues (Kumari, Head, & Marmot, 2004). The reason, according to the authors, was not simply lifestyle factors, such as being overweight. Rather, the authors pointed to differences in job satisfaction among the hierarchies of civil servants, suggesting a possible interaction between reactions to work and health.*

A study published in the Journal of Applied Psychology *reported that individuals who reported experiencing work-to-family conflict, defined as work-related impediments to meeting family-related demands, were from two to three times more likely to experience an anxiety, mood, or substance dependence disorder than their counterparts who did not report experiencing work-to-family conflict (Frone, 2000).*

These three vignettes highlight the key themes of this chapter. The first theme is that dissatisfaction with work can have significant effects on performance in the workplace. Being dissatisfied can affect the productivity of individual employees and of work groups, thereby affecting profitability and the wealth of shareholders. This effect on corporate performance is often recognized—at least tacitly—by managers, though this awareness is not always followed by effective actions. The audit of police arrests illustrates this effect and is particularly interesting because it also shows how dissatisfaction of workers can have a more generalized effect on those outside the organization. Although, as individuals, we may not appreciate receiving a citation for driving a few miles over the speed limit, we recognize that the intended goal is to enhance public safety. Job dissatisfaction can interfere with this goal, creating hazards not only for workers but for society. A similar case can be made for morale problems among airline mechanics, power plant operators, or even financial services employees who decide to retaliate by selling credit card information to fraud artists ("Bank security breach may be biggest yet," 2005).

The diabetes study and the work–family study illustrate our second theme: that work, including dissatisfaction with work, can also have important effects on employees and their families. For decades, both organizational and medical researchers have documented links between job dissatisfaction and heart disease, depression, substance abuse, and other health indicators (House, 1974; Kasl, 1974; Kornhauser, 1965; Margolis & Kroes, 1974). Dissatisfying work—and how workers cope with their dissatisfaction—can have profound effects on quality of life for employees as well as those close to them.

These vignettes also highlight a third key theme of this chapter—that the effects of work on humans can be pernicious even in the twenty-first century and in the heart of modern society. The negative effects of work are not limited to sweatshops of a prior era or far-away developing nations. Despite modern facilities, labor-saving technologies, protective laws, and increased opportunities for leisure, postindustrial work can have a substantial negative effect on the mental and physical well-being of employees.

What may have changed, at least to some extent, is how these effects occur. At one time, the primary concerns about the effects of work on individual workers pertained to accidents, occupational exposure to toxins, or wear and tear from physically demanding labor, each of which had fairly direct impacts on the well-being of workers. Primary concerns about work impacting family life were centered around whether a mother's work outside of the home had detrimental effects on her family's well being.

Although these concerns have not been eliminated from postindustrial work settings, they have been superseded by more subtle but no less insidious factors. The challenges of contemporary work are more likely to include psychologically draining interactions with supervisors, team members, or customers; worries about technological obsolescence; feeling tethered to work by cell phones, PDAs, and other technology; and having to balance the demands of family and work. The consequences of these challenges are cumulative and may not manifest in overt illness or disability for some time. In the shorter term, however, these factors are likely to influence workers' feelings about their work and how it affects their life and their families. This interest in *job satisfaction* is relatively recent and fundamentally important to our understanding of how modern work affects workers.

Thus, our focus in this chapter is more on how work affects life than vice versa. We first look at how working conditions—as mediated by job satisfaction—influence behavior at work. Our interest in this regard is looking at behavior that helps humans adapt to dissatisfying conditions, using work as a microcosm in which to study these effects. We then broaden our perspective to explore the reciprocal effects of work and non-work events and satisfaction on humans, using the bidirectional perspective of work–family research. We propose that the consequences of dissatisfaction at work, when thought of as mechanisms for adapting to dissatisfaction, may, in fact, apply more generally to life beyond work. We then close the chapter by considering how these dynamics fit into the broader context of this volume.

The Importance of Job Dissatisfaction

Affective, or emotional, reactions to work, including job (dis)satisfaction, are part of a fundamental psychological process that influences not only business outcomes but our understanding of human behavior, particularly—but not exclusively—in the workplace. Understanding how people react to work psychologically is as critical to our understanding of work as understanding how people react physiologically or as part of social groups.

Job satisfaction is generally defined as an employee's evaluation of the extent to which a job meets basic values or needs. Conversely, job dissatisfaction occurs when we do not receive what we want from our work or when the job imposes conditions that are contrary to our value preferences. As with all psychological attitudes, job satisfaction includes both cognitive (thinking, judgment) and affective (feeling, emotion-based) components. These components become important factors in how we think and feel about our lives more generally. Simply put, people who are frustrated at work are rarely upbeat about their lives in general (Tait, Padgett, & Baldwin, 1989).

What is particularly important for this discussion is that attitudes, including job dissatisfaction, also have a behavioral component. In basic approach-avoidance terms, we seek out those things that are satisfying to us and avoid those that are dissatisfying. Marketers spare no expense in utilizing this basic premise to convince us that the products they hawk will bring us satisfaction. The same premise has driven much of the early research on job satisfaction, manifested in attempts to create conditions that will enhance both job satisfaction and worker performance. Industrial psychologist Timothy Judge and his colleagues have provided a thorough review of both the theoretical logic and meta-analytic evidence for the relationship between job satisfaction and job performance, which turns out to be more complex than often thought (Judge, Thoresen, Bono, & Patton, 2001).

Our interest, in this chapter as in our research, is on the behavioral outcomes of job dissatisfaction. Psychologist Frederick Herzberg was among the first to suggest that job satisfaction is qualitatively different than job dissatisfaction, in terms of both antecedents (intrinsic factors leading to satisfaction and extrinsic factors producing dissatisfaction) and consequences (satisfaction leading to positive outcomes and dissatisfaction producing more negative outcomes) (Herzberg, Mausner, & Snyderman, 1959). Although his work had been dismissed by most organizational scholars by the late 1970s, these critiques pertained more to Herzberg's hypothesis that satisfaction and dissatisfaction had different antecedents than to the possibility that they had different consequences. Dissatisfaction is by definition unpleasant, and theories of behavior ranging from hedonism to reinforcement theory have supported the idea that undesirable states have significant motivational potential. Being dissatisfied should provide an immediate and reasonably powerful motivation to find ways to avoid the source of dissatisfaction.

When the first author of this chapter joined Charles Hulin's research group at the University of Illinois, as a graduate student in the mid-1970s, the focus in Hulin's group was on understanding how job dissatisfaction (as well as economic conditions and individual decision processes) affected employee turnover, the "ultimate" form of avoidance. One contribution to this effort was to broaden this perspective to suggest that dissatisfaction should drive not only decisions to quit but a range of dissatisfaction-driven behaviors (Rosse and Hulin, 1985). This was important because it made sense that people respond to job dissatisfaction in various ways (e.g., quitting, goofing off at work, getting frustrated and short-tempered with coworkers). This approach was also consistent with a fundamental principle that was emerging among social psychologists who were studying attitudes in the 1970s. That principle was that general attitudes (in this case, job satisfaction) are not particularly effective at predicting specific behavioral responses (e.g., quitting one's job). People act generally in accordance with their attitudes, but different people react in different ways. To see this link between general attitudes and behavior, one must have a general ("broadband") measure of behavior. Developing and refining such a measure became a goal that continues to drive our research program today.

Job Dissatisfaction and Employee Adaptation

Another contribution, initially in work with fellow graduate student Howard Miller, was to develop a more comprehensive theoretical basis for understanding the relationship between job dissatisfaction and what we have termed *adaptive behavior* (Rosse and Miller, 1984). By "adaptive," we mean that the behaviors represent an attempt, fully intentional or not, to cope with dissatisfying work events.

We began with the premise that occupational work represents a major component of modern human life. Although there is considerable variance, one might estimate that occupational work consumes at least one third of one's waking hours, and probably represents the single most extensive category of activity for most individuals. As a result, we expect that work will have a substantial effect on an individual's well-being. Of particular interest to us is that problems at work create job dissatisfaction, which, in turn, leads to responses that are intended to adapt to or cope with this dissatisfaction.

In that regard, we use the term *adaptation* slightly differently than it is used in evolutionary theory, where it conventionally refers either to a process in which natural selection results in a better fit between species and their environment or to evolved structures and behavior fashioned by selection. In this chapter, the term is used as it is in psychological research on perception, in which adaptation is thought of as an accommodation to stimuli or situations that disrupt some sort of equilibrium in an individual. In the case of job dissatisfaction, the disequilibrium is an imbalance between workers' values or desires and the demands and outcomes provided by their jobs. We started with the assumption that such disequilibrium serves as a motivating force and then sought to explore how this motivation manifests itself in the behavior of dissatisfied workers. The result is similar to evolutionary adaptation, though the timeline and the process are more similar to that of niche construction (as described by Laland and Brown in chapter 5) than to natural selection.

In early experimental work in the late 1970s, we asked people to describe what they did at work when they were dissatisfied. This produced long lists of specific—sometimes graphic—descriptions of behavior. But we were not really interested in predicting highly specific behaviors, whether that behavior was quitting one's job or negotiating with a coworker to swap days off. Rather, we were interested in studying general behavioral tendencies that resulted from general affective reactions to the job (i.e., job dissatisfaction). This required developing broadband behavioral measures that would correspond to the broadband measures of job satisfaction. That search began by exploring existing models that seemed relevant.

In 1970, social scientist Albert Hirschman applied his economic model of dissatisfaction to the topic of employee dissatisfaction (Hirschman, 1970). His economic theory concluded that when customers are dissatisfied with the product or service of a company, their primary choices include Exit (obtaining the product or service else-where), Voice (making the organization aware of their dissatisfaction and allowing an opportunity to fix the problem), or Loyalty (waiting patiently for things to improve). (Because many factors in behavioral research are named with words that are also used in common speech, researchers often follow the practice of capitalizing the title of a factor (e.g., Exit) to differentiate it from the more general use of the same word, a practice that we follow in this chapter.) According to Hirschman, these same principles would hold true when employees are dissatisfied with their employers. In an employ-ment setting, *Exit* consists of leaving the job or organization, *Voice* implies actively trying to resolve the problem, and *Loyalty* suggests passively waiting for conditions to improve.

Hirschman's notion of Exit resonated strongly with organizational researchers who had been studying the phenomenon generally referred to as *employee withdrawal*, which suggested that turnover, absenteeism, employee lateness, and other related behaviors represented alternative ways of "escaping" from unpleasant work. More recently, Exit has been conceptualized as more extreme forms of withdrawal (i.e., quitting, transferring, retirement) that are distinct from more transitory withdrawal behaviors such as absenteeism and lateness (which are included in a different behav-ioral category) (Hanisch & Hulin, 1991).

Voice was a more novel concept for researchers interested in organizational behav-ior. That may be surprising, because it seems self-evident that trying to find ways to improve one's lot would be a basic method of coping with dissatisfaction. Indeed, it is at the core of most theories of how people cope with stress. One reason that it had received less attention among organizational researchers may be that such research had been dominated by the interests and values of managers and firms, who may be inclined to see some Voice responses—such as proposing new ways of doing things or forming or joining a labor union—as threatening the status quo. The perspective of our research group was that the adaptive value of a behavior should be viewed from the perspective of the actor. From the employee's frame of reference, some forms of Voice that may seem to conflict with a firm's interests may be an adaptive, or func-tional, way of responding to dissatisfying working conditions. Even so, most Voice responses represent problem-solving solutions that are likely to be beneficial to both employees and their firm.

More than a decade after Hirschman's theory, the concept of *Neglect* was introduced as another behavioral expression of dissatisfaction. This occurred almost simultane-ously in Rosse's dissertation (Rosse, 1983) and in work by social psychologist Caryl Rusbult and her colleagues in their exploration of responses to dissatisfaction in roman-tic relationships (Rusbult, Farrell, Rogers, & Mainous, 1988). Neglect, also referred to as Avoidance, involves careless or neglectful performance, shirking of responsibilities, persistent unjustified absences or lateness, and various forms of loafing on the job.

Retaliation has also been documented as a distinct behavioral reaction to employee dissatisfaction (Glomb, 1999; Rosse, 1983). Retaliatory behaviors may serve to rectify perceived inequities in the workplace. An example might be an employee who feels that she is not being paid what she is worth. One way to redress that inequity might be to steal from her employer, whether overtly or through more subtle forms such as claiming more hours than she actually worked. This type of *equity-enhancing retaliation* is presumed to be relatively rational (in the literal sense of being calculated) and may be particularly related to the cognitive/evaluative dimension of job dissatisfaction. An affective, or emotional, component of job satisfaction may lead to a different form of retaliatory behavior that may not on its face seem adaptive. This may manifest in such actions as yelling or cursing at coworkers or customers, sabotaging the work of others, or even physical assaults. Behaviors such as these may reflect *cathartic retaliation,* behavior driven by a desire to "even the score" with an employer, supervisor, or coworker that is the source of dissatisfaction. We regard retaliatory behaviors, particularly those of a cathartic nature, as being driven by the adaptation process because of their motivation, not because their long-term consequences will necessarily be functional—even for the actor. In this way, this use of the term *adaptive* raises issues similar to those raised in the discussion of maladaptive behaviors by Laland and Brown in chapter 5 and by Levin, Laland, and Saturay in chapter 8.

When we first began to conduct research on this model, it quickly became apparent that, in many cases, there is no behavioral response to dissatisfying working conditions. Some dissatisfied individuals do not try to change the situation, they do not exit the situation, and even their normal work performance seems to be relatively stable. In attempting to explain these situations, we more recently have identified an additional response category that we refer to as *Capitulation* to reflect the nonreactive nature of these responses (Levin & Rosse, 2001; Miller & Rosse, 2002). We now hypothesize that this response category includes subcategories and that only the most extreme subcategory fully encompasses the notion of giving up that the Capitulation label implies.

The first subcategory of Capitulation is called *Loyalty*, borrowed from Hirschman's original model. Loyalty involves patiently waiting for the problem to resolve; this represents a temporary form of adaptation with the expectation that things will eventually get better without active intervention by the individual. Undesired events are virtually guaranteed to occur at some frequency in even the best of jobs, and the resulting sense of dissatisfaction often fades with the passage of time, even as quickly as overnight (Judge & Ilies, 2004). Responding with a degree of patience is an efficient way to conserve psychological, physical, and social resources for those situations that are more serious or persistent.

Adjusting expectations, the second category of Capitulation, involves a reevaluation by the employee of the situation and of his or her expectations of what a job should require or provide. The disappointment of discovering that a job is not as desirable as one expected is a common cause of turnover, as well as other adaptive behaviors, in new employees. It is also a common experience for employees experiencing a major change in working conditions, such as surviving a layoff or other restructuring. If initial expectations were unrealistically high, readjusting them lower may be a functional way of coping with this reality shock.

Finally, a third kind of Capitulation is *Disengagement.* The employee has evaluated the dissatisfying situation, decided that no course of action is going to produce more desirable circumstances, and responds by deciding to stay in his or her current position. This is literally giving up and often shows up as employee burnout: a syndrome in which people disengage from their work, depersonalize clients and coworkers, and become emotionally exhausted (Maslach, Schaufeli, & Leiter, 2001). This category of response may not be truly adaptive, as it may result in significant physical and mental health consequences (Rosse & Hulin, 1985). It has much in common with the responses to inescapable stressors and learned helplessness discussed in chapter 9 of this volume.

Our work on this taxonomy has progressed both deductively, using prior models, and inductively, based on a series of interviews and surveys in which employees have been asked to describe how they respond to dissatisfaction. This work continues, but we now have reasonable confidence that the behavioral families are relatively distinct from one another and encompass fairly well the various ways in which people respond to dissatisfaction. Based on that foundation, we have also done experimental work to explore the patterns of relationships between job dissatisfaction and these behavioral families and to begin to explore the reasons that people choose different strategies for adapting to dissatisfaction.

One experimental question has to do with the relative frequency of the different responses. Rosse & Saturay (2004) found that the most common responses to dissatisfying work events were Exit and Disengagement. But what was even more interesting was that all of the categories of adaptation were used by a substantial number of employees, supporting the perspective that adaptation is a multifaceted phenomenon.

This variance in response across employees raises the question of why these differences in response exist. One common perspective is that differences in work behavior are attributable to individual characteristics, with more desirable employees being loyal, dependable, and willing to persevere in the face of challenge, while other employees are quick to respond with "deviant" behaviors. There is some evidence that people react differently to the experience of dissatisfaction, although these individual differences tend to have less to do with personality or values (nearly all of which were statistically unrelated to differences in behavioral responses) than with differences in how constructively people manage conflict (Rosse & Saturay, 2004). Prior history, particularly whether prior attempts to adapt were successful, may also be important for predicting how people respond to dissatisfaction. Voice, or problem-solving, was more common among those who were newer to their jobs, whereas Neglect was more likely among those who were longer term employees. One possible explanation is that those who become frustrated with futile attempts to improve their work then resort to Neglect or Exit.

Two steps were needed in order to establish that these behavioral families are adaptive: first, that job dissatisfaction results in these behaviors, as confirmed above; second, that engaging in these behaviors provides a means of coping with the dissatisfaction. Rosse (1983) explored this by investigating the health consequences of not using an adaptive mechanism when dissatisfied. He found elevated levels of self-reported symptoms of mental and physical distress but no effects on blood lipids among employees who were dissatisfied but reported using none of the adaptive behavior alternatives. More recently, Rosse & Saturay (2004) found that employees who reported engaging in adaptive behaviors after a dissatisfying event subsequently reported feeling less dissatisfied, suggesting that the behaviors may indeed be psychologically adaptive.

The Role of Non-Work Factors

One of the limitations of much of this research on satisfaction and adaptation is that its focus has been limited to occupational work. That is, the sources of dissatisfaction have been limited to characteristics of a person's job or employing organization, and the exploration of adaptive responses has focused principally on job-related behaviors. However, just as positive feelings resulting from job satisfaction may spill over into the home and family life of individuals, job dissatisfaction may engender negative feelings and behaviors in a person's life outside the workplace (Wilensky, 1960).

The spillover between work and non-work domains is bidirectional. That is, time constraints, stress, and behavioral restrictions can spill over from family life into work life and vice versa. Sources of dissatisfaction and stress rooted outside the work role may exacerbate feelings of dissatisfaction with work. Positive spillover is possible as well; employees may use leisure activities to compensate for constraints and deprivations experienced at work.

One area in which it is interesting to extend the discussion of satisfaction and adaptation out of the strict sphere of the workplace is that of work–life conflict and how it impacts the performance and behavior of employees at work. As mentioned, occupational work typically consumes one third or more of an individual's waking hours, with the remaining time devoted to personal responsibilities and occasional leisure activities. For many workers, personal responsibilities include the care and maintenance of a home and/or family, including such tasks as childcare, elder care, or spousal/life partner obligations, in addition to cooking, cleaning, household maintenance, and shopping. Although all of these (and other) activities could be included under the definition of work used in this volume, we will follow the convention of referring to a clash between occupational work responsibilities and familial responsibilities as *work–family conflict.*

Work–family conflict is formally defined by Kahn and his colleagues as a form of interrole conflict in which one's obligations in a particular role interfere with one's ability to fulfill obligations in another role (Kahn, Wolfe, Snoek, & Rosenthal, 1964). These conflicts may be time based, in which time is the constraining factor; strain based, in which stress from one role spills over into another role; or behavior based, in which behavior required in one role is incompatible with behavior required in the other role (Greenhaus & Beutell, 1985).

We hypothesize that individuals attempt to adapt to the experience of work–family conflict using the same mechanisms used to adapt to dissatisfaction with their work. However, the reciprocal nature of the relationship between work life and home life makes this process dynamic. For example, satisfaction with the ways in which one's employer accommodates familial obligations often translates into greater satisfaction on the family front, because more time is available to meet these familial obligations. Less stress is then likely to spill over from work into home life, and behaviors are likely to be more compatible across the two domains. Likewise, satisfaction with family and the division of labor within the household often translates into a smoother integration of family commitments with work responsibilities.

In the mid-twentieth century, the American workforce was composed primarily of married men with wives who did not work outside of the home and who held primary caretaking responsibility for the couple's children. In 1950, for example, women constituted less than 30% of the total American workforce and had a labor force participation rate of less than 34% (Kutscher, 1993). In the twenty-first century, however, fewer than 15% of American families fit this description (Parasuraman & Greenhaus, 2002). Today, the workforce comprises many more dual-earning couples, working mothers, single parents, and representatives from other forms of nontraditional families. Women now make up almost one half of the current American workforce, and current labor force participation rates for women are more than double what they were in 1950 (Chao, 2001). While some of these emerging forms of families create the impetus for a more equal division of labor inside the home, others, such as single-parent families, undoubtedly create more conflict between the two domains and exacerbate the cyclical nature of their relationship. Is it possible to do well at a secure and interesting job, keep a decent house, and do well by your spouse, friends, and children all at the same time?

A study by the Families and Work Institute helps to illustrate the evolving interaction between work and family. During the technological boom of the mid-1990s, many employees appeared to place higher priority on career success than on family.

In fact, members of the baby boom generation are almost twice as likely to place a primary focus on work activities as members of subsequent generations (Families and Work Institute, 2004). The increase in the average age of first-time mothers seems to support this notion, presuming that women delayed childbearing until they were better established in their careers. Furthermore, the study found that individuals who place an equal emphasis on their families and their work and those who place a primary emphasis on their family responsibilities are more satisfied with their jobs and their lives in general than those individuals whose primary focus is their occupational work (Families and Work Institute, 2004).

These findings have two implications. The first is that there has been a shift over the last few decades in the workforce's relative emphasis on work and family. This shift in focus toward the family, which has boosted the job and home life satisfaction of employees, also provides an interesting example of adaptation at a more aggregate level than the individual. As stated previously, dissatisfaction results from the experience of disequilibrium, such as that created by conflicting demands placed on individuals by their work and their personal lives. By adopting behaviors that minimize the source of dissatisfaction and equalize the disequilibrium, dissatisfaction is reduced. In this case, the adaptive behavior is shifting one's primary focus from the domain of work to the domain of life outside of work. The Family and Work Institute study suggests that there may have been a generational adaptation in response to dissatisfaction experienced by previous generations whose primary focus was on work life. The implications of a hypothesized generational adaptation extend well beyond the management of organizations and employees. It suggests that when a large enough constituency adopts a change in behavior, or an adaptation, to a specific source of dissatisfaction, it can result in a societal change that, in turn, might induce a change in organizations, such as the introduction of childcare benefits and family leave policies. Such changes ultimately affect our occupational work lives, making them more reflective of present-day attitudes about work.

Understanding the nature of the relationship between work and non-work life is important for the three reasons illustrated above. First, one's level of satisfaction or dissatisfaction at work has consequences not only for the employer and the employee but also for the employee's family members. This manifests in the form of the employee's mental, physical, and emotional well-being and the potential for stress and counterproductive behaviors to spill over into an employee's personal life. The positive or negative effects of this spillover are without bounds, making the understanding of the relationship between work and personal domains critical. Second, dissatisfaction due to work–family conflict can heighten feelings of dissatisfaction at work. This may result in the employee engaging in the adaptive behaviors described above, some of which can be detrimental to organizational outcomes. Lastly, employee attempts at adapting to their dissatisfaction with opportunities to integrate work and family can have profound effects on society, particularly when there is a common source of dissatisfaction among a large percentage of employees. Organizational responses to employee adaptations can alter the organizational environment and help to move toward eliminating certain sources of dissatisfaction altogether.

Implications of Dissatisfaction and Adaptation for Understanding Modern Work

The most commonly accepted definitions of *work* among management scholars indicate that work involves purposeful activity intended to create something of value. This definition has a number of implications for what has been discussed in this chapter and throughout this volume. For one, "creating value" conjures a very functional view of work, wherein effort and attention are focused almost exclusively on the production of valued goods or services. Such rational/economic views of work provide little insight into such behaviors as goofing off, gossiping about the organization or coworkers, being uncooperative toward coworkers, or being absent for a mental health day. These behaviors can be understood, however, when viewed as responses to dissatisfying work. That is not to suggest that job dissatisfaction is the sole cause of these behaviors, but there is now a significant body of research indicating that job (dis)satisfaction is one of the key predictors of workplace job performance and counterproductive behaviors, broadly defined.

This definition of work also implies that we leave our homes to go to a place that we conventionally call *work*, in order to perform a series of tasks that create value for the organization and for society. At the end of the day (or shift), we then leave that world to return to the world of home and family. As Maier and Levin note in chapter 9, this artificial separation of work from the rest of life does not make sense from a physiological perspective. It is even less accurate in contemporary knowledge-based work, due to changes in technology (e.g., mobile phones and laptop computers), work design (e.g., telecommuting), employment contracts (e.g., the preponderance of salaried employees and the demise of labor unions), and social changes (e.g., increasing trends for both partners in a family to be employed out of the home).

These trends suggest a much tighter coupling of work and life, so that spillover of events and feelings from work to life and vice versa should be expected. As noted above, the work-life satisfaction research suggests that while these effects are bidirectional, problems at work may have a disproportionately strong effect on family roles and on life satisfaction more broadly. One interesting question is the extent to which work and non-work roles may also play a coping function. A recent study by Tim Judge and his colleague indicates that employees often bring a bad mood home with them but that this affective cloud is largely dissipated by the time they return to work the next day (Judge & Ilies, 2004). Their findings are consistent with research showing that sleep may play a major role in this recuperative function. It is often suggested that vacation time, hobby activities, and involvement with family members may also play a role in dissipating the effects of work stress. Focused research on these hypotheses might be a useful in advancing our understanding of how work affects life.

We, in this chapter, and Maier and Levin in chapter 9, have both noted that stresses and hassles at work can have significant effects on employees' mental and physical health, as well as their levels of satisfaction. It is particularly interesting that research on both animal physiology and human adaptation and stress responses has identified personal control as a key moderator of the stress–strain relationship. Having a sense of control over one's fate seems to be a powerful factor in reducing dysfunctional consequences of stress. The work on reactions to controllable and uncontrollable stressors described in chapter 9 shows how the presence of control, when detected in the cortex, eliminates physiological reactions to uncontrollable stressors that otherwise proceed by way of the brainstem in rats. In humans, workplace research suggests that employees with a greater sense of control may also be more likely to choose constructive responses to job dissatisfaction. Extending these findings might lead to the hypothesis that issues of control are particularly salient in the workplace. (Indeed, many employees might find similarities between their work and the electrified grids on which rats were studied!) Perhaps effective management systems are those that can create a sense of empowerment that can generalize to life more broadly? Similarly, we might expect that ineffective management systems can have a robust effect on reducing overall perceptions of control and efficacy, thus having an effect that extends well beyond the workplace.

References

Bank security breach may be biggest yet (2005, May 23). *CNN Money*. Retrieved from http://money.cnn.com/2005/05/23/news/fortune500/bank_info/

Chao, E. L. (2001). *Report on the American workforce*. Washington, D.C: U. S. Department of Labor.

Families and Work Institute. (2004). *Generation and gender in the workplace*. New York, NY: American Business Collaboration.

Frone, M. R. (2000). Work family conflict and employee psychiatric disorders: The national comorbidity study. *Journal of Applied Psychology, 85*, 888–895.

Glomb, T. M. (1999). *Workplace aggression: Antecedents, behavioral components, and consequences*. Paper presented at the annual meeting of the American Psychological Society, Denver, CO.

Greenhaus, J. H., & Beutell, N. J. (1985). Sources of conflict between work and family roles. *Academy of Management Review, 10*, 76–88.

Hanisch, K., & Hulin, C. (1991). General attitudes and organizational withdrawal: An evaluation of a causal model. *Journal of Vocational Behavior, 39*, 110–128.

Herzberg, F., Mausner, B., & Snyderman, B. B. (1959). *The motivation to work*. New York: John Wiley & Sons.

Hirschman, A. O. (1970). *Exit, voice, and loyalty: Responses to decline in firms, organizations and states*. Cambridge, MA: Harvard University Press.

House, J. (1974). Occupational stress and coronary heart disease: A review and theoretical integration. *Journal of Health and Social Behavior, 15*, 12–27.

Judge, T., Thoresen, C. J., Bono, J. E., & Patton, G. K. (2001). The job satisfaction-job performance relationship: A quantitative and qualitative review. *Psychological Bulletin, 127*, 376–407.

Judge, T. A., & Ilies, R. (2004). Affect and job satisfaction: A study of their relationship at work and at home. *Journal of Applied Psychology, 89*(4), 661–673.

Kahn, R., Wolfe, D., Snoek, J., & Rosenthal, R. (1964). *Organizational stress: Studies in role conflict and ambiguity*. New York: Wiley and Sons.

Kasl, S. (1974). Work and mental health. In J. O'Toole (Ed.), *Work and the quality of life* (pp. 171–196). Cambridge, MA: MIT Press.

Kilzer, L. (2005, August 20). Traffic tickets dive. *Rocky Mountain News*, p. 4A.

Kornhauser, A. (1965). *Mental health of the industrial worker*. New York: Wiley and Sons.

Kumari, M., Head, J., & Marmot, M. (2004). Prospective study of social and other risk factors for incidence of Type 2 diabetes in the Whitehall II study. *Archives of Internal Medicine, 164*, 1873–1880.

Kutscher, R. E. (1993). Historical trends, 1950–1992, and current uncertainties. *Monthly Labor Review, 116*(11), 3–10.

Levin, R. A., & Rosse, J. (2001). *Talent flow: A strategic approach to keeping good employees, helping them grow, and letting them go*. San Francisco, CA: Jossey-Bass/Wiley.

Margolis, B., & Kroes, W. (1974). Work and the health of man. In J. O'Toole (Ed.), *Work and the quality of life* (pp. 134–144). Cambridge, MA: MIT Press.

Maslach, C., Schaufeli, W. B., & Leiter, M. P. (2001). Job burnout. *Annual Review of Psychology, 52*, 397–422.

Miller, H., & Rosse, J. (2002). Emotional reserve and adaptation to job dissatisfaction. In J. M. Brett & F. Drasgow (Eds.), *The psychology of work: Theoretically based empirical research* (pp. 205–231). Hillsdale, NJ: Erlbaum.

Parasuraman, S., & Greenhaus, J. H. (2002). Toward reducing some critical gaps in work-family research. *Human Resource Management Review, 12*, 299–312.

Rosse, J. (1983). *Employee withdrawal and adaptation: An expanded framework* (Unpublished doctoral dissertation). University of Illinois, Urbana-Champaign.

Rosse, J., & Hulin, C. L. (1985). Adaptation to work: An analysis of employee health, withdrawal and change. *Organizational Behavior and Human Decision Processes, 36*, 324–347.

Rosse, J., & Miller, H. E. (1984). Relationship between absenteeism and other employee behaviors. In P. S. Goodman, R. S. Atkin & Associates (Eds.), *Absenteeism: New approaches to understanding, measuring, and managing employee absence.* (pp. 194–228). San Francisco: Jossey-Bass.

Rosse, J., & Saturay, S. (2004). *Individual differences in adaptation to work dissatisfaction.* Paper presented at the annual meeting of the Western Academy of Management, Fairbanks, AK.

Rusbult, C. E., Farrell, D., Rogers, G., & Mainous, A. G. (1988). Impact of exchange variables on exit, voice, loyalty, and neglect: An integrative model of responses to declining job satisfaction. *Academy of Management Journal, 31*, 599–627.

Tait, M., Padgett, M. Y., & Baldwin, T. T. (1989). Job and life satisfaction: A reevaluation of the strength of the relationship and gender effects as a function of the date of the study. *Journal of Applied Psychology, 74*, 502–507.

Wilensky, H. L. (1960). Work, careers, and social integration. *International Social Science Journal, 12*, 543–560.

This chapter presents an intriguing proposition that has been in development since the outset of the book. This chapter, by Robert Levin, Kevin Laland, and Stacy Saturay, starts with energetic trade-offs similar to those described in earlier chapters. It proposes that these trade-offs in an organism might be divided between energy conservation and energy utilization, labeled in this chapter as "working energy/take-home energy trade-offs." Based on exquisitely developed mechanisms for energy allocations in animals, the chapter posits that all animals, including humans, will be sensitive to the balance in these energy allocations and that humans' sensitivity to this balance allows researchers to explore many contemporary workplace issues in terms of the effect of energy allocations. In turn, this allows Levin, Laland, and Saturay to use energetic and evolutionary approaches described earlier, with nuance provided by the hypothesis of human sensitivity to changes in the balance inherent to ongoing energetic trade-offs. The subjects of job dissatisfaction and work–life balance are then reexplored from the perspective of these hypothesized working energy/take-home energy trade-offs.

8   Do Energy Allocations Affect Work Performance? The Working Energy/Take-Home Energy Trade-off Hypothesis

Robert Levin, Kevin Laland, and Stacy Saturay

Foundations for Energetic Trade-offs and Implications for Work in Organisms

Complex knowledge-based work in modern organizations seems a world removed from the work of foraging for energy sources that is observed in other organisms, in ancestral humans, and in some hunter-gatherer and foraging societies today. Indeed, the development from preindustrial to industrial to postindustrial work has been depicted as freeing both workers and organizations from physical and biological life-history constraints (Rabinbach, 1990), such as the continuous need to forage for energy sources simply to survive in a harsh environment (Prosser, 1986). The ongoing effects on knowledge workers in modern organizations of biological processes, such as competing requirements for physical energy in individuals, the topic of this chapter, are rarely considered when assessing worker performance or when identifying and explaining the problems and conflicts that can occur when knowledge work is performed. Moreover, as postindustrial work has increasingly emphasized cognitive work over physical work, frameworks utilized by researchers and in organizations for understanding individual performance have also tended to emphasize psychological, behavioral, or social foundations, rather than physical and biological foundations. The ubiquity of cognitive postindustrial work can lead both scholars and managers to view workers almost as disembodied brains.

Our approach to understanding individual human work performance in postindustrial settings in this chapter and in this volume seeks to reintroduce biological and physical foundations in a manner that does not challenge the centrality of established premises of the importance of cognitive, behavioral, and social factors in how human work gets done. Instead, we seek here to explore the implications of the following link between physiology, behavior, and work: that because humans, like all organisms, have biological systems with highly developed and finely-tuned sensitivity to how they allocate energy reserves, they, like all organisms, endeavor to maintain a critical balance in energy allocation to meet specific alternative demands. We describe this balance as the *working energy/take-home energy trade-off*.

Human workers are undoubtedly complex social creatures, but they are also biological beings. Some of the capacities and limitations that humans encounter in performing work today are linked to fundamentals underlying all life on Earth and human biological processes in particular. These fundamentals affect aspects of all human work, both physical and cognitive. These critical limits and influences on behavior, grounded in biological fundamentals common to all forms of life, have an impact on levels of energy expenditure at work in ways that are not always apparent from current perspectives.

We begin by exploring developments in evolutionary and biological research that may be useful tools for understanding these trade-offs and their effects on work in organisms in general and in humans in particular. In chapter 5 of this volume, Laland and Brown present in detail developments in understanding biological and evolutionary influences on human behavior that are also relevant to our discussion here, including the development of the niche-construction perspective and the rise of human behavioral ecology. Most directly related to the discussion here is the development of energetic, rather than reproductive, concepts of biological fitness, a development that is crucial to applying evolutionary constructs to explorations of contemporary human work in useful ways. Each of these areas is discussed in detail by Laland and Brown (2002) and Laland and Brown in chapter 5.

Turner's (2000) physiological definition of fitness is a useful development with which to understand adaptation and behavior in contemporary humans. Emphasizing the capacity of many organisms to control and regulate energy fluxes both within and outside of their bodies, Turner suggests that biological fitness (i.e., the capacity to survive and reproduce) can be profitably understood as the ability of an organism to gather and deploy energy. We suggest that this energetically based fitness measure potentially provides a relevant currency for fitness that is useful even in modern postindustrial societies. It also provides a connection to the various energetic discussions in this volume (e.g., chapters 1, 2, and 4). This concept of fitness is used in this chapter to explore the potential relationships between energetic trade-offs and modern human behavior related to the workplace.

Exploring Energetic Trade-offs

We have, over some time, been exploring the effects of trade-offs between conflicting requirements for energy in humans and other organisms (Levin & Laland, 2003; Levin, Laland, & Grant, 2004). Our research has come to focus on a trade-off between two basic energy requirements in all organisms: the requirement for sufficient energy for short-term demands, including survival, which we will refer to functionally as *working energy* and the requirement for energy surpluses essential for such critical but less immediate functions as longer term survival, safety in the face of environmental stress, growth, reproduction, and rearing—surpluses we will refer to as *take-home energy*. The conflicts and trade-offs between the uses of energy for short-term moment-to-moment needs and longer term surpluses, between utilization and storage, are fundamental to many underlying human biological systems (some examples of which are provided below). Modern work can, on the one hand, create rigid and extreme demands for time and energy and, on the other hand, create opportunities and demands for behavioral flexibility that vastly exceed and differ from the kinds of conditions under which human physiological and neurobiological capabilities for processing energy and information evolved.

Our use of the term *energy* is not intended metaphorically. From a biological perspective, physical and chemical energy is required, as noted throughout this volume, for all forms of biological function, from maintaining body temperature to walking, performing physical work, reproducing, and raising offspring. This is true for cognitive as well as manual labor (Laughlin, van Steveninck, & Anderson, 1998; see also chapter 2). Brains are known to be energetically expensive organs (Aiello & Wheeler, 1995). The energetic demands for cortical computation, for example, are high enough to restrict the fraction of the brain's cortical computational capacity that can be used at any one time (Lennie, 2003; Niven & Laughlin, 2008). Moreover, even though cognition and information transfer may be less visibly apparent activities than physical activities, their energy demands are met by precisely the same metabolic work processes within the cell and the organism used to meet the demands of physical activity (Harold, 1986; see also chapter 1).

Competing requirements for the limited supply of physical energy that any organism can obtain from its environment can generate conflicting natural selection pressures. An organism that spent the entire amount of energy it obtained from the environment merely to survive would not produce any viable offspring, and, if it were a general characteristic, its species would die out. Conversely, an organism that devoted too great a proportion of its resources to building up energy surpluses would struggle to survive in the short term, and over the longer term might exhibit suboptimal fitness. Thus, mechanisms that allow organisms to manage the balance between the competing demands for working energy and take-home energy have very high adaptive value and are more likely to be evolutionarily preserved.

It follows that, over evolutionary time, what becomes important to a given organism is both the absolute levels of energy it can attain and the maintenance of a working energy/take-home energy balance within an acceptable range. That balance should provide for survival, safety, growth and development, reproduction, and rearing—all within some feasible solution limited by the total amount of energy obtained. Humans, like other animals, have several biological and neurophysiological systems that work in just this manner, as described below.

Meanings of *Work* in Energetic Trade-offs

Physicists have a precise usage of the term *work*, which on the surface seems a world away from the meaning of this term in postindustrial cognitive work. Work (in this physical or thermodynamic sense) is the processes that organisms engage in that allow them to exchange energy with their environments, to channel energy through their bodies, and to create orderliness in the process. Laland and Brown (chapter 5) develop the argument that thermodynamic and vocational senses of *work* are, in fact, related. They maintain that "work (vocational) is that subset of work (thermodynamic) that involves activities that function to accrue an energy surplus (i.e., for which the gross energy accrued is greater than the working energy cost)" (p. 127)

Applying Turner's (2000) concept of energetic fitness to the energetic trade-offs proposed in this chapter suggests that organisms' activities, such as foraging, have been favored by natural selection because they typically result in net energy accrual and the production of order in biological systems. Such activities benefit organisms by recouping the energy cost of the activities themselves and the cost of other energy-requiring physiological processes throughout the duration of the activity (working energy) but also generate energy surpluses or stores (take-home energy). These energy stores have many uses; one is that they can allow for, or facilitate, future reproduction. Another crucial use is to provide for safety in the face of fluctuating environmental conditions. We have also elsewhere analyzed the trend of these surpluses over time and proposed that take-home energy surpluses must ultimately be the source of the energy and time required for the use of "information," in the sense of Laughlin's work in chapter 2 (Levin, Laland, and Grant, 2004).

Such net-energy-accruing activities should also be regarded as work in the vocational sense. When humans go to the workplace today, they trade their exertions, as noted in chapter 5, for institutionalized tokens, such as money, that can be converted into resources—such as food, heating, or protection from the elements—that ultimately provide them with energy in various forms.

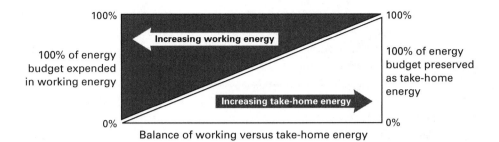

Figure 8.1
Relationship of working energy to take-home energy.
Original illustration by Alan Blackwell.

Ultimately, both the physiological success of organisms and the reproductive success of a population depend on their being able to accrue energy in a manner that minimizes the working-energy cost and maximizes the surplus that can be allocated to reproduction by efficiently translating the take-home energy into viable descendants that are themselves fit, given the constraints necessary to stay alive. From the perspective of Turner's (2000) physiological definition of fitness, this energy surplus bears a direct correspondence to biological fitness. Given a finite limit on the energy that a human can accrue, greater energy expended in work, in the vocational sense, means less take-home energy (figure 8.1).

Conversely, greater energy can only be allocated to take-home energy (a) by increasing the efficiency of work (such that, for a given unit of working-energy cost, greater net surplus energy is accrued) or (b) by reducing the working-energy cost (Levin, Laland, and Grant, 2004). Given this inherent trade-off between possible energy demands, all organisms, including humans, should have evolved adaptations that function to maintain an optimal balance (see chapter 3) of energy allocation to working energy and take-home energy budgets. Organisms have, in fact, evolved such adaptations; exquisitely tuned adaptations to changes in energy balance are among the earliest adaptations apparent in the simplest organisms (Atkinson, 1965; Harold, 1986).

Examples of Biological Mechanisms Producing Energetic Trade-offs
The general understanding that organisms have developed well-established mechanisms for responding to and balancing immediate and longer term demands for energy has a long lineage in physiology (e.g., Alexander, 1999; Hochachka & Somero, 2002; Schmidt-Nielsen, 1972; Thompson, 1917 [1961]). Two examples of such adaptations, in which the mechanisms of energetic trade-offs are particularly clear and are discussed in other contexts elsewhere in this volume, are given below.

Energy Regulation by Adenosine Balance   As noted in chapter 1, a bacterial organism regulates its own activity from moment to moment by the ratio of adenosine triphosphate (ATP), adenosine diphosphate (ADP), and adenosine monophosphate (AMP) (Atkinson, 1965; Harold, 1986). This regulation and change in activity level goes on several thousand times each second. A higher relative concentration of ATP equates to greater take-home energy and capacity for working energy. A higher relative concentration of ADP and AMP indicates a greater expenditure of working energy and a depletion of take-home energy reserves, because ADP and inorganic phosphate must be synthesized together to form ATP, using the intense energy demands from chemiosmosis noted in chapter 1.

Responses to Stressors   A second example from this volume, at a different timescale and level of complexity, is the energetic response to stressors, described in chapter 9. In the framework presented here, responses to uncontrollable stress tend to represent a shift from working energy toward increased take-home energy, as energy is conserved, for instance, when organisms are under attack. Responses utilizing the bidirectional brain–immune circuit represent shifts in both directions: Preparing a fight–flight response would represent a shift from take-home energy to working energy, while "sickness behavior" would represent a shift from working energy to take-home energy, conserving energy to fight infection and preserving energetic stores to allow reduced foraging. The insulin response in mammals is also an example of a complex system that regulates the bidirectional shifts between working energy and take-home energy on an ongoing basis, regulating a balance in the face of dynamic activity of the organism.

These examples, and many more possible, illustrate the existence of effective physiological mechanisms for detecting and responding to trade-offs in working and take-home energy in humans.

Energetic Trade-offs and Contemporary Human Work: Exploring Energetic Effects on Satisfaction, Dissatisfaction, and Work Performance

One implication of numerous, widespread, evolutionarily ancient mechanisms producing a working energy/take-home energy tradeoff is that all organisms—not simply humans or other complex organisms—are predicted to be sensitive to the balance between working energy demands and take-home energy levels and to react on an ongoing basis to changes in this balance. This implies that humans, like other organisms, will act in ways that are consistent with a tendency toward seeking and maintaining a balance in energy allocation between the conflicting demands of working and take-home energy within a physiologically acceptable range.

Cognitively based work and the behavioral flexibility of modern humans in postindustrial society, far from making individual humans immune to such trade-offs, instead allow expenditures of time and energy that may regularly expose humans to extreme energy demands. Modern organizations, constructed work environments, and knowledge-based work demand and draw on a much higher degree of behavioral flexibility for humans than exists for any other organism or than have likely existed for humans in other times and settings. Nonetheless, we maintain that this behavioral flexibility will be subject to the effects of, and often challenge, fundamental biological constraints imposed by the need to maintain an acceptable working energy/take-home energy balance.

In this section, we explore how some aspects of human behavior in the workplace that have appeared puzzling can be analyzed in a new light when regarded as contemporary manifestations of long-established mechanisms for adaptive energy allocation and for protecting an adaptive energetic balance.

For this purpose, we have picked an easily accessible example to see how it may be informed by the working energy/take-home energy tradeoff hypothesis; namely, the relationship between job satisfaction, dissatisfaction, and performance described in chapter 7. We begin this exploration by utilizing the working energy/take-home energy trade-off to conceptualize in energetic terms (a) the potential effects of demands by an employer for increased time or energy expenditure by an employee and (b) the predicted impact of these demands on dissatisfaction and on performance.

The adaptive behavior model of responses to job dissatisfaction is reviewed in detail in chapter 7. For our purposes in this chapter, the relationship of adaptive behavior to energetic trade-offs is twofold. First, combining the working energy/take-home energy trade-off model with the adaptive behavior model provides a way for us to explore linkages between energetic demands, dissatisfaction, and potential effects on performance. We suggest that "adaptive behavior" is not only behaviorally adaptive (i.e., effective, functional) but also, in the terms of the working energy/take-home energy trade-off hypothesis and of Turner (2000), biologically adaptive (i.e., such behavior can function to increase take-home energy in the face of the demands on the worker made by contemporary organizations). Second, in making this analysis, we are able to make specific conceptual and empirical predictions about the effects of different sources of dissatisfaction on such areas as performance, based on humans' sensitivities to these energetic trade-offs. For instance, the analysis that follows might lead researchers to consider when biological factors could trigger "adaptive behavior" (e.g., neglect or retaliation) in the workplace.

To begin, we use the working energy/take-home energy trade-off hypothesis to consider the effects of demands on a worker to spend greater amounts of time at work. These demands could be induced or coerced by the employer, through peer pressure by a work group in the manner of concertive control (Barker, 1993), or through an employee's own volition. Assume that, prior to these demands, the worker has been working in conditions under which his or her working energy/take-home energy demands are in balance. In that case, work under these new conditions would change or distort the optimal working energy/take-home energy balance, as demands for take-home energy go increasingly unfilled. This shift in energy allocation away from that balance would then, in turn, lead to dissatisfaction.

This linkage of physiological disequilibrium with dissatisfaction can be conceptualized as occurring in either of two ways. The first possible pathway is purely physiological: Manifold mechanisms operate to keep an organism in energetic balance. Disturbing these balances produces physiological discomfort which, endured over a period of time in a work setting, produces dissatisfaction. A second possibility is that, over evolutionary time, behavioral responses have evolved that produce changes that are experienced as changes in behavior and motivation, including dissatisfaction, as a direct response to the change in working energy/take-home energy balance. Chapter 9 provides an extended example of a parallel effect of physiology on affect and behavior in its discussion of sickness behavior and of the behavioral reactions to stressors.

The predicted relationship between take-home energy and satisfaction is represented in the schematic in figure 8.2.

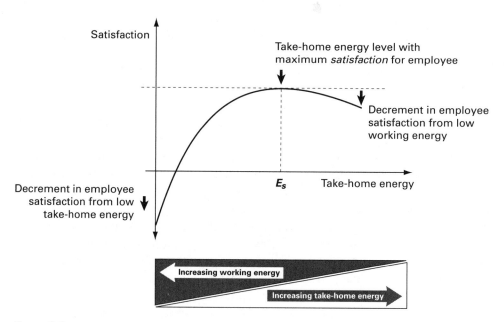

Figure 8.2
Predicted relationship of satisfaction to take-home energy.
Illustration credit: Alan Blackwell.

Satisfaction (and especially dissatisfaction) is represented in figure 8.2 as affected by the extent to which an individual has struck a suitable balance in working and take-home energy allocations. Satisfaction would therefore be expected, all else being equal, to peak at intermediary levels of take-home energy that coincide with this balance. Moving away from this balance would be expected to lead to dissatisfaction, whether through investing too much or too little energy in work. The relationship between take-home energy and satisfaction is depicted as curvilinear. With excessive levels of energy expended at work, take-home energy demands would increasingly go unmet, and workers would increasingly leave work exhausted and drained: Satisfaction correspondingly decreases. (As in chapter 9, this chapter does not differentiate between the sources of stress between work and life, as these are not differentiable on a physiological basis, though they may be differentiable for other purposes. The same approach is used with regard to sources of satisfaction and dissatisfaction.)

Conversely, at some point, for most individuals, increased allocations to take-home energy (e.g., through loafing, absences, or leaving work for a range of recreational activities (Rosse and Miller, 1984)) interfere with necessary requirements to hold productive employment and earn a satisfactory living. To the extent that sanctions (or guilt) occur, satisfaction should again decrease. A key prediction is that maximal satisfaction will not occur with maximal take-home energy but rather at some intermediary level of take-home energy. In figure 8.2, the level of take-home energy that provides maximal satisfaction to an employee is shown as E_S.

Now consider the relationship of take-home energy to performance (figure 8.3). Performance in the workplace is expected to correspond to the amount of energy that employees dedicate to their jobs; that is, to working-energy demands. However, just as with satisfaction, the relationship between performance and take-home energy is postulated to be curvilinear. When employee allocations to take-home energy are very high and available working energy is, therefore, lowered, work performance is also likely to be lowered, because employees have devoted relatively lowered effort to their work. As allocations to take-home energy are reduced, working energy increases and so does performance, all else being equal. This prediction contrasts with an energy-cost-free idealization of postindustrial information-based work, which would in contrast suggest that performance should continue to rise monotonically with increased allocations to working energy. This idealization is represented by the dashed line rising to the left in figure 8.3.

Several factors contribute to the precipitous drop in performance indicated as well in figure 8.3 when take-home energy becomes increasingly restricted by increased demands for working energy. First, we posit that upsetting the desired working energy/take-home energy balance triggers dissatisfaction. Dissatisfaction, in turn, results in an increased frequency of counterproductive behavior and, thus, across a workforce, would result in various performance reductions (Levin & Rosse, 2001). If adaptive

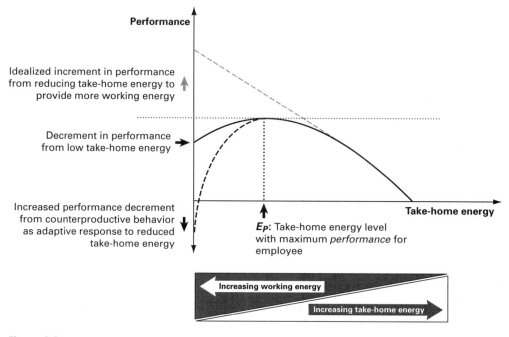

Figure 8.3
Predicted relationship of performance to take-home energy.
Illustration credit: Alan Blackwell.

behavior is also energetically adaptive, then we hypothesize that these changes occur, in part, because humans possess physiological and psychological adaptations that increase the likelihood they will behave in ways that protect the working energy/ take-home energy balance and, with it, vital levels of take-home energy.

At the extreme depicted at the far left of figure 8.3, performance can be not only disrupted but truly negative, in that extreme counterproductive behavior by individuals reduces the global productivity of the organization (Levin & Rosse, 2001). In addition, especially in circumstances in which this extreme demand is self-imposed, excessive working-energy demands begin to result in physical exhaustion, sleep deprivation, and burnout (which themselves may be physiological and behavioral adaptations acting in ways that prevent individual organisms from investing too much energy in work), further reducing performance. Collectively, these factors result in the predicted decrement in performance represented by the dashed line falling to the left in figure 8.3.

Thus, we do not predict that maximal job performance occurs when employers attempt to maximize the working energy obtained from employees. Rather, optimal performance from the perspective of the employer occurs at an intermediary level of take-home energy, which in figure 8.3 is represented by E_P. E_P represents the level of take-home energy at which the employee would deliver maximal performance and at which the employer derives the maximum benefit from the employee's activities. (This intermediary level is also consistent with the implications of design centering in multiple-objective optimization presented by Lightner in chapter 3.)

Note that the term *performance* denotes the actual work performance of an individual worker. (The figures in this chapter would thus represent the average actual work performance of a worker at a given point in time.) In real-life settings, the issues for both managers and organizational researchers of effectively measuring performance—in particular "negative" performance or performance related to counterproductive behavior—are considerable: Doing so accurately is very difficult, and doing so at all is too often neglected (Campbell, McCloy, Oppler, & Sager, 1993; Levin & Rosse, 2001; Levin & Zickar, 2002). Those significant measurement problems are, however, separate from the underlying relationships we seek to explore here. Indeed, one of the unexpected results of this analysis is that it may provide a way to tackle one of the unexplained mysteries of organizational research—the inability of research studies in workplaces to consistently establish a relationship between satisfaction and performance, despite the fact that one is widely expected (Iaffaldano & Muchinsky, 1985).

Now we can examine the predicted relationship of working energy and take-home energy to both satisfaction and performance simultaneously. Because take-home energy is equal to total energy minus working energy, with total energy assumed to be constant, working energy and take-home energy can be plotted on the same axis (figure 8.4).

We have two reasons for expecting the functional relationship between satisfaction and performance depicted in figure 8.4. First, the adaptive behavior model demonstrates empirical relationships between satisfaction (particularly dissatisfaction) and apparently counterproductive behavior, with the latter negatively affecting performance.

Second, and more fundamentally, the nature of the working energy/take-home energy trade-off, illustrated in figure 8.1, suggests that satisfaction and performance constrain each other. In figure 8.4, the level of take-home energy at which the employee would deliver maximal performance (E_P) is assumed to be less than or equal to the level of take-home energy that provides maximal satisfaction to an employee (E_S). This is because, while employees will seek to protect their own take-home energy, employers will not be similarly sensitive to preserving or enhancing conservation of employee take-home energy. Quite the contrary: Up to a point, the interests of employers in enhanced performance will not be served by allowing employees all the take-home energy they wish but instead by obtaining more working energy from their employees, to the extent that doing so increases performance.

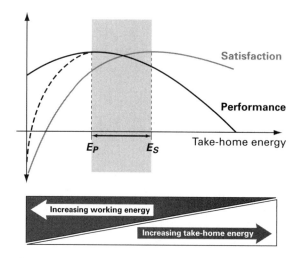

Figure 8.4
Combined relationships of satisfaction and performance to take-home energy.
Illustration credit: Alan Blackwell.

Elsewhere, we have developed a formal proof that the trade-off being postulated will occur in the intersection of any two concave-down (or n-shaped) curves, regardless of their relative heights, so long as E_P is, as postulated, less than or equal to E_S (Levin, Laland, and Grant, 2004). Thus, the existence of the trade-off and of the resulting *zone of conflict* discussed below are not dependent on the absolute or relative heights of the curves; that is, to the absolute or relative magnitudes of working energy or take-home energy demands. The magnitude—but not the existence—of the trade-offs and of the zone of conflict are related to the absolute and relative magnitudes of working energy and take-home energy demands and to the magnitude of the relationship between dissatisfaction and performance.

The working energy/take-home energy trade-off implies that an employer's efforts to increase working-energy demands beyond E_P would represent a false economy. Rather than obtaining increased performance from the increased demands for working energy, employers are likely to experience decreased performance mediated, for example, through dissatisfaction and counterproductive behavior, because a critical point has been passed in the working energy/take-home energy trade-off.

These boundaries, although not visible or apparent, can function in the same manner as hard boundaries on resources, such as factory capacity or the availability of cash. Particularly in employment situations in which employers are in a position to set the rules, the resulting practices and policies can easily drift into and across these boundaries. Suggesting that a salesperson who has sold 50 units can sell 60 units may be reasonable, but it does not follow that the same salesperson can, therefore, sell 70 units without passing a critical working energy/take-home energy boundary. These energetic boundaries need to become apparent to researchers, executives, managers, and workers. These energetic boundaries and working-energy/take-home energy trade-offs are therefore examples of the performance envelopes introduced and noted throughout this volume.

Figure 8.4 illustrates three distinct regions of take-home energy; for each, the relationship between satisfaction and performance differs, represented by the zones with levels of take-home energy below E_P, above E_S, and between E_P and E_S. A long-term, mutually beneficial employment relationship is feasible only in the middle region between E_P and E_S. That is because, in the region in which levels of take-home energy are less than E_P, it is overtly in the interests of both employees and employers to reduce working-energy demands. Failing to do so would tend to result in dissatisfaction, which is not in the individual interest of an employee, and in the negative performance effects shown in figure 8.3, which is not in the interest of the employer.

Similarly, situations in which take-home energy exceeds E_S are inherently unstable. A feather-bedding employee might assume that regularly calling in sick will provide increased take-home energy. But when performance drops below what is required by the employer because of insufficient working-energy allocations, then sanctions, reduced income, disruption to career development, or job loss will occur, and satisfaction will be lowered. In such a case, neither the needs of the employee nor the needs of the employer are being met.

Note that, below E_P and above E_S, increased satisfaction is related to increased performance, and decreased satisfaction is related to decreased performance. In contrast, within the region between E_P and E_S, direct trade-offs occur between performance and satisfaction. Moreover, within this region, employers may seek greater performance through greater working-energy demands, moving toward E_P, while employees may seek greater satisfaction through greater take-home energy demands, moving toward E_S.

Thus, in this critical central region, increased performance occurs at the expense of satisfaction, and increased satisfaction occurs at the expense of performance. We could, therefore, refer to this region, where employer–employment relations are stable, yet where satisfaction–performance trade-offs are inherent, as the *zone of conflict*. This term is meant to suggest that trade-offs between working energy and take-home energy and between satisfaction and performance are inherent in a stable long-term employment relationship.

The establishment and maintenance of a stable balance in this critical region, and the level of take-home energy at which a balance is struck, would depend on the absolute and relative strengths of demands for working energy and take-home energy and on the relative power of employers and employees. Reducing absolute levels of demands for both working energy and take-home energy does not eliminate conflict but instead appears to broaden the range of mutually acceptable options and may reduce the magnitude of conflict. Researchers, employers, and employees might, therefore, profitably seek to understand how to broaden ranges of options within the zone of conflict and how to ameliorate but not eliminate the effects of these trade-offs. The region might, therefore, equally be referred to as the *region of compromise.*

While these results contradict the common notion that job satisfaction is positively related to job performance, they may also, in fact, help explain the empirical, meta-analytic finding that, over the entire range of satisfaction, the two constructs are only weakly related (Iaffaldano & Muchinsky, 1985). The expectation of a positive relationship between performance and satisfaction at the extremes and a negative relationship at intermediate values is a testable prediction of the working energy/take-home energy trade-off hypothesis. Thus, it may be fruitful for researchers to investigate relationships between time at work, energetic expenditure at work, and satisfaction, dissatisfaction, and performance. For instance, researchers could administer a questionnaire in which they ask workers to judge whether they feel that they invest too much, too little, or approximately suitable levels of time and energy at work relative to non-work life. Were the same individuals' levels of satisfaction (or dissatisfaction) and performance to also be assessed (noting the inherent difficulty of assessing performance referred to above), we would expect positive relationships between satisfaction and actual work performance in the "too much" and "too little" groups and negative relationships in the "suitable levels" group. (Alternatively, take-home energy could be assessed quantitatively by the covariate of "time not spent at work.")

The working energy/take-home energy trade-off hypothesis leads us to suggest more broadly that satisfaction–performance trade-offs may be the norm within relatively stable employment relationships. This approach, therefore, differs from assuming that either the trade-offs or the presence of dissatisfaction are aberrant and can be made to go away. Situations in which increasing satisfaction improves performance may, in fact, signify aberrant or unstable conditions—situations in which demands on workers (or on employers) are at a level that cannot be sustained. Instead, satisfaction–performance trade-offs and the existence of the zone of conflict appear to be inherent in the energetic constraints of work and the conditions required for a stable, long-term employment relationship.

Work–Life Conflict and Working Energy/Take-Home Energy Trade-offs

We can now apply the postulated relationship between satisfaction, performance, and working energy/take-home energy trade-offs to exploring how this approach might apply to contemporary issues in the area of *work–life balance* or *work–life conflict*—two revealingly paired terms. When considering the balance between working energy and take-home energy that many employees are attempting to achieve, it is necessary, for example, to consider the energetic requirements that individuals often face with regard to family responsibilities. Because working energy refers to those energy demands an individual faces for his or her own immediate "survival" needs, energy needed to fulfill family responsibilities, such as child rearing, is classified here as a take-home-energy demand. In some respects, working energy demands today are largely the same as they have always been. The human body requires generally the same amounts of food, water, and rest, and external energy (e.g., for heat) for survival as it always has, although we are arguably less in touch with these requirements than were our ancestors because of physical and conceptual changes in our workplaces and because foraging, as described earlier, takes place through obtaining tokens (i.e., money) that are used for food rather than by gathering food itself.

At first glance, these changes in modern-day employment circumstances might seem to lend themselves to a more convenient way of balancing demands between work, in a vocational sense, and family. Having the flexibility to work during any hour of the day, rather than just during daylight hours, or to purchase whatever food products we desire rather than relying on what we can kill or forage, seems that it would allow more energy for family responsibilities.

However, this is just one possibility, because with increased behavioral and social flexibility, the energy savings could just as easily be instead allocated to the workplace. Indeed, technological advances leading to putative labor-saving devices at work and away from work, combined with timeless social pressures in which individuals are judged according to wealth or status, have created a climate in which many individuals dedicate increasing, rather than decreasing, amounts of energy to the workplace, in particular to obtain the devices that can save labor at home (Hunnicutt, 1988). Changes in work circumstances, such as the introduction of technology, artificial lighting, and heating, tend to allow knowledge workers (Drucker, 1967) to expend more hours and working energy at their jobs, squeezing the surplus energy available to maintain family responsibilities.

Modern-day knowledge workers may have largely lost track of energetic concerns that were obvious and vital to workers of earlier times, but energetic mechanisms have not ceased to operate. Technological advances, increased demand to acquire labor-saving devices, and social pressures can create a "ratchet effect" in which allocations to working energy increase inexorably. Labor-saving devices are, in this context, only completely "labor-saving" if they are freely acquired and do not require workers to expend extra energy to purchase the device. They would be "net labor saving" only if there is energy saved after the additional labor required to acquire them. The introduction of the washing machine is the device that most clearly changed the labor balance in the family and the home (Hunnicutt, 1988). It is less clear that the stainless-steel-exterior version of the same washing machine (with computerized controls) can make any claim to further reducing labor, let alone after accounting for the increased labor required for its increased cost.

With the addition of each labor-saving device, then, greater amounts of energy can be devoted to work. Contrary to many early concerns about impacts of automation and computer technology (Hayes, 2009; Leontief, 1952), technological advances at the workplace have not necessarily led to increased leisure time (Whaples, 2001).

In this context, a volume recently published by Pecchi and Pigga (2008) is of interest. In it, contributions from noted economists provide their views on why economist John Maynard Keynes predicted correctly in 1930 that our individual wealth would rise substantially over the 70 years following, but predicted incorrectly that our working hours would fall to about 15 hours per week. Like other observers, Keynes thought that lives filled with technology would also be filled with the social challenge of how to deal with a population that would work so little (Hayes, 2009; Pecchi & Pigga, 2008).

The economists provide a variety of explanations (Pecchi & Pigga, 2008); many of the economists suggest that increases in labor-saving technology do not sate humans' desire to acquire technologies and, thus, the increased leisure predicted by Keynes has been turned into increased requirements for earnings to acquire material goods. These explanations are consistent with the earlier, carefully documented work of Hunnicutt (1988) showing that a consumer economy largely depends on just such a shift and that American workers shifted over time from striving for a shorter work week to striving to acquire costly consumer goods, such as the automobile and the washing machine. Whaples (2001) has documented that, although measuring working hours is inherently difficult, working hours for U.S. workers, while falling steadily beginning in the late 1800s, have not fallen substantially over the period in which Keynes predicted they would fall by nearly two thirds.

In the terms of the working energy/take-home energy trade-off hypothesis, the longer term effect of this ratcheting process, at least among workers who can set their work hours with flexibility, would be to push individuals to allocate energy to work in a way that approaches or passes the feasible boundary of the zone of conflict indicated by E_P. Beyond E_P, there is no employer–employee conflict. However, our model would predict, and some empirical research indicates, the presence of substantial work–life conflict and distress (e.g., Brett & Stroh, 2003).

Thus, widespread cognitively-driven behavioral flexibility with regard to work can create conflicts between biological work and cultural work that violate the working energy/take-home energy balance, in turn leading to or exacerbating what has come to be called work–family conflict or work–life conflict. Work–family conflict is defined as a form of "interrole" conflict in which demands from one role (e.g., work) inflict upon demands from another role (e.g., family) (Greenhaus & Beutell, 1985; Kahn, Wolfe, Quinn, Snoek, & Rosenthal, 1964). Work–family conflict has definite consequences for employee satisfaction, including a persistent, negative relationship between the experience of work–family conflict and work and life satisfaction (Kossek & Ozeki, 1998). Work–family conflict has been positively related to family distress for blue-collar workers (Frone, Russell, & Cooper, 1992) and to unpleasant mood spillover from work to family and family to work (Williams & Alliger, 1994).

Work–family research with regard to employee satisfaction typically espouses the spillover hypothesis described in chapter 7 to explain the relationship between the two domains, with *work-to-family spillover* meaning that work responsibilities invade family time; *family-to-work spillover* means that family responsibilities impose on time that should be spent working. The working energy/take-home energy trade-off hypothesis provides a more energetically or biologically based explanation of the importance of such spillover effects. Moreover, it should be possible to make specific qualitative predictions about when spillover will occur, the direction spillover will take, and about causal factors: Specifically, when workers are investing too much energy at work, there should be a "squeeze" on take-home energy, leading to problems at home, while conversely, where family responsibilities require excessive time and effort, there should be a squeeze on working energy.

We conclude with a note about the "problem of leisure," one raised by economists, as we have noted, and by various researchers who discussed our early thought experiments with us. Does the allocation of energy to activities known as "leisure" provide any sort of challenge to the working energy/take-home energy tradeoff framework? No, it does not: The allocation to leisure is simply an allocation of energy to take-home energy. It is not inherently maladaptive in the sense of Turner's (2000) concept of energetic fitness. The question at hand is whether the volume of allocation of take-home energy to leisure activities creates an adaptive or maladaptive working energy/take-home energy balance: Many human societies over time have provided regulated opportunities for moderate amounts of leisure time (e.g., holidays, vacations), while sanctioning volumes of leisure that might jeopardize sufficient allocations to working energy (e.g., Aesop's fable of the grasshopper and the ant). Likewise, it will not be news to most readers that, in a family setting, increased allocations of take-home energy to leisure activities by a parent, at the neglect of allocating take-home energy to various family responsibilities, have been known to cause interpersonal conflict between human couples on at least more than one occasion.

Summary of Implications

We close by summarizing the implications of working energy/take-home energy trade-offs:

1. The working energy/take-home energy trade-off approach provides a broad framework with which to explore ways in which biological processes, particularly those

related to the ongoing utilization and storage of energy, can affect the behavior of modern humans, including in the workplace.

2. Working energy/take-home energy trade-offs are based on innumerable biological mechanisms that help organisms allocate energy between immediate and longer term needs.

3. These mechanisms are vital to an organism's survival, often date to the earliest times of life, and are sensitive to ongoing fluctuations. Animals are, therefore, sensitive to shifts in the working energy/take-home energy balance. These physiological mechanisms and trade-offs operate in present-day humans.

4. Working energy/take-home energy trade-offs can be useful for analyzing and seeking to understand areas of contemporary work, such as work performance, employee satisfaction and dissatisfaction, and the relationship between these parameters and the balance between work and non-work life.

5. The common assumption that performance is likely to be positively linearly related to time or energy devoted to work turns out to be problematic. Instead, we would expect curvilinear relationships between both satisfaction and performance, with each peaking at intermediary levels of working and take-home energy.

6. The common assumption that satisfaction and performance are positively related cannot be assumed to hold across the full range of levels of working energy. On the contrary, working energy/take-home energy trade-offs suggest that satisfaction and performance will be negatively related for most ongoing, stable employer–employee relationships, and this negative relationship is inherent within a zone of conflict.

7. Where satisfaction and performance are positively related, we would expect employer-to-employee relationships to be unstable.

8. Efforts on the parts of employers to achieve marginal performance gains by elevating working-energy demands beyond a critical limit (E_P) would represent a false economy, as these elevated demands come to impinge take-home energy levels and elevate dissatisfaction.

9. We make a number of empirical predictions that suggest that it would be fruitful for researchers to investigate relationships between time at work, perceived energetic expenditure at work, and satisfaction, dissatisfaction, and performance.

10. The working energy/take-home energy trade-off hypothesis leads us to anticipate a tension between time and energy dedicated to work and home, which may frequently manifest itself in work–life conflict.

11. These trade-offs may provide a biological explanation for aspects of the existence and importance of spillover effects with regard to satisfaction between work and non-work life, and for their bidirectional nature.

12. The long-term effect of a ratcheting process of devoting working energy to obtain "labor-saving" devices or services, freeing more time and energy to devote to work, would push individuals to allocate energy to work that approaches and

passes the boundary of the zone of conflict indicated by E_P. Past this point, there is no employer–employee conflict; however, we would expect substantial work–life conflict as a result.

Working energy/take-home energy trade-offs thus function as one of the performance envelopes identified throughout this book. While some performance envelopes function at an absolute level, delineating feasible from infeasible work, working energy/take-home energy trade-offs function somewhat differently to delineate between solutions that may be feasible in the short term from those that may be feasible in the long term, helping the organism achieve a necessary balance between these two. In this way, these trade-offs also mediate between the physiological and the psychological, affecting energetics, behavior, and work in ways both nuanced and fundamental.

Acknowledgments

Christina De La Rocha, John Hoffecker, Simon Laughlin, Michael Lightner, and John Odling-Smee made contributions to some of the earliest versions of the working energy/take-home energy trade-off model. David Allen, David Grant, and Joseph Rosse each contributed content and comments to earlier versions of this manuscript. Linda Argote and Greg Northcraft made helpful extended comments on another form of this manuscript, which led to the chapter being developed in its current form. We thank Alan Blackwell for his illustrations and for the discussions which accompanied them, which sharpened our thinking.

References

Aiello, L. C., & Wheeler, P. (1995). The expensive-tissue hypothesis: The brain and the digestive system in human and primate evolution. *Current Anthropology, 36*(2), 199–221.

Alexander, R. M. (1999). *Energy for animal life*. Oxford: Oxford University Press.

Atkinson, D. E. (1965). Biological feedback control at the molecular level. *Science, 150*(3698), 851–857.

Barker, J. R. (1993). Tightening the iron cage: Concertive control in self-managing teams. *Administrative Science Quarterly, 38*, 408–437.

Beis, I., & Newsholme, E. A. (1975). The contents of adenine nucleotides, phosphagens and some glycolytic intermediates in resting muscles from vertebrates and invertebrates. *Biochemical Journal, 152*, 23–32.

Brett, J., & Stroh, L. (2003). Working 61 plus hours a week: Why do managers do it? *Journal of Applied Psychology, 88*(1), 67–78.

Campbell, J. P., McCloy, R. A., Oppler, S. H., & Sager, C. E. (1993). A theory of performance. In N. Schmitt, W. C. Borman, & Associates. (Eds.), *Personnel selection in organizations* (pp. 35–70). San Francisco: Jossey-Bass.

Drucker, P. F. (1967). *The effective executive*. New York: Harper and Row.

Frone, M. R., Russell, M., & Cooper, M. L. (1992). Antecedents and outcomes of work family conflict: Testing a model of the work-family interface. *Journal of Applied Psychology, 77*(1), 65–78.

Greenhaus, J. H., & Beutell, N. J. (1985). Sources of conflict between work and family roles. *Academy of Management Review, 10*(1), 76–88.

Harold, F. M. (1986). *The vital force: A study of bioenergetics*. New York: W.H. Freeman.

Hayes, B. (2009). Automation on the job. *American Scientist, 97*(1), 10–14.

Hochachka, P. W., & Somero, G. N. (2002). *Biochemical adaptation: Mechanism and process in physiological evolution*. New York: Oxford University Press.

Hunnicutt, B. K. (1988). *Work without end: Abandoning shorter hours for the right to work*. Philadelphia: Temple University Press.

Iaffaldano, M.T., & Muchinsky, P.M. (1985). Job satisfaction and performance: A meta-analysis. *Psychological Bulletin, 97*, 251–273.

Kahn, R. L., Wolfe, D. M., Quinn, R. P., Snoek, J. D., & Rosenthal, R. A. (1964). *Organizational stress: Studies in role conflict and ambiguity*. New York: John Wiley & Sons.

Kossek, E. E., & Ozeki, C. (1998). Work-family conflict, policies, and the job-life satisfaction relationship: A review and directions for organizational behavior-human resources research. *Journal of Applied Psychology, 83*(2), 139–149.

Laland, K. N., & Brown, G. R. (2002). *Sense and nonsense: Evolutionary perspectives on human behaviour*. New York: Oxford University Press.

Laughlin, S. B., van Steveninck, R. R. D., & Anderson, J. C. (1998). The metabolic cost of neural information. *Nature Neuroscience, 1*(1), 36–41.

Lennie, P. (2003). The cost of cortical computation. *Current Biology, 13*, 493–497.

Leontief, W. (1952). Machines and man. *Scientific American, 187*(3), 150–160.

Levin, R. A., & Laland, K. N. (2003, February). The working energy/take-home energy hypothesis. In R. A. Levin, A. C. Bekoff, J. G. Rosse, and C. L. De La Rocha (Chairs), *Putting energy and information to work in living systems*. Symposium conducted at the annual meeting of the American Association for the Advancement of Science, Denver, CO.

Levin, R. A., Laland, K. N., & Grant, D. R. (2004). How do organisms respond to competing energetic requirements? The working energy/take-home energy hypothesis [Abstract]. *Integrative and Comparative Biology, 44*(6), 719.

Levin, R. A., & Rosse, J. G. (2001). *Talent flow: A strategic approach to keeping good employees, helping them grow, and letting them go.* San Francisco: Jossey-Bass/Wiley.

Levin, R. A., & Zickar, M. J. (2002). Investigating self-presentation, lies, and bullshit: Understanding faking and its effects on selection decisions using theory, field research, and simulation. In J. Brett & F. Drasgow (Eds.), *The psychology of work: Theoretically-based empirical research* (pp. 253–276). Mahwah, NJ: Erlbaum.

Niven, J. E., & Laughlin, S. B. (2008). Energy limitation as a selective pressure on the evolution of sensory systems. *Journal of Experimental Biology, 211*(11), 1792–1803.

Pecchi, L., & Piga, G. (2008). *Revisiting Keynes: Economic possibilities for our grandchildren.* Cambridge, MA: MIT Press.

Prosser, C. L. (1986). *Adaptational biology: From molecules to organisms.* New York: Wiley.

Rabinbach, A. (1990). *The human motor: Energy, fatigue, and the origins of modernity.* New York: Basic Books.

Rosse, J. G., & Miller, H. E. (1984). Relationship between absenteeism and other employee behaviors. In P. S. Goodman, R. S. Atkin, & Associates. (Eds.), *Absenteeism: New approaches to understanding, measuring, and managing employee absence* (pp. 194–228). San Francisco: Jossey-Bass.

Schmidt-Nielsen, K. (1972). *How animals work.* Cambridge: Cambridge University Press.

Thompson, D. A. (1917 [1961]). *On growth and form* (Abridged edition, edited by John Tyler Bonner). Cambridge: Cambridge University Press.

Turner, J. S. (2000). *The extended organism: The physiology of animal-built structures.* Cambridge, MA: Harvard University Press.

Whaples, R. (2001). Hours of work in U.S. history. In R. Whaples (Ed.), *EH.Net Encyclopedia.* Retrieved from http://eh.net/encyclopedia/article/whaples.work.hours.us.

Williams, K. J., & Alliger, G. M. (1994). Role stressors, mood spillover, and perceptions of work-family conflict in employed parents. *Academy of Management Journal, 37,* 837–868.

This chapter, by Steven Maier and Robert Levin, explores the phenomenon commonly referred to as stress *in relation to work. Maier and Levin begin by placing stress in the broader context of environmental challenges. The responses to these challenges show that a single challenge or stressor can trigger many different kinds of biological systems and many different kinds of responses. This framework challenges us to rethink such concepts as "stress at work." Whether a stressor happens to occur at work or at home may, for example, be far less important than whether it is controllable or uncontrollable. Maier and Levin explore the neuroscience research behind this distinction and the implications of this distinction in depth. In addition, Maier and Levin show how responses to stressors have come to utilize mechanisms that originally evolved to fight infection—mechanisms that produce significant changes in energy conservation or mobilization and in the capacity to perform work. They also develop an unexpected implication of these findings: Individuals have significant differences in their own performance from moment to moment and day to day—differences that are often overlooked both in psychological theory and in workplace practice but that can be the foundation for a more productive understanding of the ways in which work meets life.*

9 Responding to the Challenges of the Environment: Stressors, the Brain, and Work

Steven Maier and Robert Levin

Work and *stress* are two words that seem to be paired together frequently. Understandably, in our day-to-day work lives, we are often concerned with issues of job stress and life stress and how these seem to affect each other. In this chapter, we want to start by taking a step back from such concerns—as important as they are—by examining stress and work from research that starts with a different perspective.

In doing so, we can place a common notion such as "job stress" into a much broader, better defined, and more useful framework. This framework places stress into a broad range of challenges to which animals must respond, referring to all of these as *environmental challenges*. Environmental challenges would include an attack by a predator, an infection, the loss of a loved one or a job, or sitting in a committee room or a personnel office facing the loss of one's position.

Animals have survived over time because a wide variety of responses to environmental challenges have arisen, developed, evolved, and been conserved. A single environmental challenge can trigger more than one of these responses under different conditions; different kinds of environmental challenges can trigger the same response. The nature of many responses to environmental challenges is that they affect an animal's capability to perform work. We may say, "Work is causing me a lot of stress"; in this chapter, we describe, in fact, how *responses to stressors* directly affect capabilities to perform work.

We discuss responses of animals (including humans) to stressors and to other environmental challenges, such as infection. We explore the effects these responses have on animals and see how profoundly these responses affect the ways in which animals perform work. We explain fundamental questions raised by research on responses to stressors and environmental challenges: Research on responses to uncontrollable stressors has recently resulted in conclusions about the importance of the absence or presence of *control* over stressors (Amat et al., 2005; Maier, Amat, Baratta, Paul, & Watkins, 2006). Research on infection has shown the existence of *bidirectional control* between brain and immune system (Maier & Watkins, 1998). Bidirectional control is also, through evolutionary adaptation, involved in responses to stressors in the absence of infection. In turn, such control changes our view of the effect of the environment on the brain and on behavior, in a human workplace and elsewhere. (Our discussion of research on uncontrollable stress and on bidirectional control in this chapter is based on the reviews by Maier et al. (2006) and by Maier and Watkins (1998), where more specific information and references may be found.)

This perspective on stress and work, in fact, regards the distinction between "work stress" and "life stress" as largely irrelevant to the effect on the organism. *Responses* to stressors, however, have profound impacts—for instance, in the brain and in the immune system—on behavior and the performance of work and can be triggered in any setting and persist in time through different settings. We humans are not two organisms, shedding one organism when we enter the workplace to don another and reversing the process when we leave each evening. Instead, we are one organism, with many distinct biological systems. The responses triggered by stressors in any of these systems have profound effects on how we perform at work. Those effects not only affect us as individuals but may ripple through organizations and economies as the cumulative, largely unrecognized, unpredictable effects of innumerable individual stress reactions.

Throughout our exploration in this chapter, responses to stressors repeatedly appear to inhibit the successful execution of work. Because, at face value, this appears to be maladaptive, it is worth noting that the physiological responses to stressors present in modern humans have aspects that are (or at least at some time were) evolutionarily adaptive. They also have aspects related to the fact that the brain and immune systems of animals did not emerge fully formed but evolved over at least the last 500 million years from other less complex biological chemicals, structures, and processes. These ancient starting points have both shaped and constrained the evolution of brains and the immune system. For this reason, some of the molecules involved in the responses to stressors play multiple roles within an organism, and the same molecule can play different roles within different responses to stressors. In other words, different systems, which emerged at different points in evolutionary time, activate different physiologically adaptive responses to stressors.

Thus, this exploration is both evolutionary, in the general meaning of that term, and is also energetic. These two features are often related. The responses to stressors described here involve changes in the way that animals utilize energy, the mechanisms for which have developed as evolutionary adaptations. The energetic nature of responses to stressors is consistent with the kinds of performance envelopes and energetic trade-offs described in chapters 1, 2, 3, 4, and 8 and constitute one more set of the energetic characteristics of animal and human work.

Some key concepts and terminology in this chapter are intended to create more useful distinctions than is conveyed by the term *stress,* which tends to be used in a wide range of general ways. It is more helpful to refer instead to a *stressor,* which exists in the environment, as well as to a *response to a stressor,* occurring in the organism. A stressor, for example, might be an attack by a predator. Or it might be environmental conditions that present the possibility of an attack by a predator. Different stressors might evoke different responses by triggering different systems or processes. Stressors are one kind of threat to survival, safety, or well-being that animals must successfully confront in their environment. In this chapter, then, we place stressors within the broader category of *environmental challenges.*

For the purposes of this chapter, another important variety of environmental challenge is *infection.* The aspect of infection that is of interest is that infectious agents are detected and responded to by the immune system. Infection/immune challenge is virtually omnipresent (and invisible) in natural animal and human environments. Responses to infection are important because they profoundly affect how animals and humans can do work, and because one of the coordinated systems for responding to stressors has come, through evolutionary processes, to utilize pathways in the brain and immune system that initially developed as responses to infection, rather than to stressors. We will focus, therefore, in this chapter on various responses to environmental challenges encountered by all animals, including humans.

Uncontrollable Stressors and the Effects of Control

What is an *uncontrollable stressor*? An uncontrollable stressor is one that is uninfluenced by the behavior of the subject. Uncontrollable stress is part of an animal's experience in nature and is also a frequent part of the human experience and of the human workplace.

Discovering the Effects of Uncontrollability and Controllability

It is possible to determine experimentally whether the outcomes or responses to a stressor are a consequence of the stressor's uncontrollability/controllability or are reactions to the stressor itself: In a paradigmatic experimental arrangement, laboratory rats might be presented with a series of mild electric shocks to the tail in one of two conditions. The rats in the first condition (referred to as the escape condition) can turn off (and thus escape) each of the mild tail shocks, by pressing a bar or panel or by turning a wheel. The rats in the second condition (referred to as the yoked condition) cannot turn the shock off—here, turning the wheel or pressing the lever has no consequence. Each shock starts at the same instant for the yoked subject as for the escape subject and terminates for both subjects whenever the escape subject responds. The level of controllability of the stressor is the only difference between the two conditions, as the length of each shock experienced by the rats in the second condition is the same as for the rats in the first condition. An experiment of this design produces two related sets of results. First, animals that have control over the stressor continue to respond to shock by wheel-turning to turn off the shock. However, animals that do not have control over their environment stop trying to turn off the shock. Second, the persistence of this effect into new environments indicated the importance of a distinct kind of response to uncontrollable stress, which was initially referred to as *learned helplessness* (Maier & Seligman, 1976; Seligman & Maier, 1967).

When animals are next presented with a problem in a different environment, one in which both groups of animals are now able to escape shock—for instance, by jumping over a barrier—the two groups of animals show a continued and persistent divergence: Animals that had previously experienced control over the stressor (in the escape condition) quickly learn to escape the shock in the new condition. They consistently jump the barrier, either to escape the shock or to avoid it in the first place.

However, animals that had earlier experienced the same volume of the initial stressor but did not have control over the stressor (in the yoked condition) respond differently. They do not take action to avoid or terminate the shock in the new environment, even though they are now free to control their exposure to the stressor. Moreover, even if they avoid the shock inadvertently by jumping the barrier, the animals apparently do not learn from this experience (Amat et al., 2005; Maier et al., 2006; Seligman & Maier, 1967). What questions can these results help us ask with respect to uncontrollable and controllable stressors and the effects of control?

Pathways toward Understanding Effects of Stressors and Control

In the remainder of this portion of the chapter, we examine integrative research on animals' behavioral responses to uncontrollable stressors, and to the presence of control, and the neurological, physiological, and biological foundations of these responses. These foundations can show us whether and how these relationships are fundamental to animals' responses to stressors, and what implications can be drawn for the evolution of distinct responses to stressors and to their effects on our present-day lives and work.

Other research streams have focused on assessing how humans, in particular, respond to the presence or absence of control in their environment and whether humans respond similarly to experimental animals. Such research indicates this often to be the case: We present two examples, both from naturally occurring settings, of research on whether humans respond similarly to experimental animals in the face of uncontrollable stress.

First, psychologists studying wellness in a geriatric-care setting augmented the care delivered to residents on two floors of the facility, providing the same augmentation in two different ways. On the first of these floors, the augmentations were administered by the staff and without control by the residents. For example, each resident received a new plant, which was cared for by the staff. On the second floor, the residents were involved with and controlled the changes—for instance, residents cared for the new plant they received. They also generally received encouragement to take more control over and responsibility for their own care. Not only did both the physical and mental health of the second group improve dramatically, according to physicians' records, but the 18-month mortality rate of the second group dropped from 25% to 15%; the mortality rate of the first group rose from 25% to 30% (Langer & Rodin, 1976; Peterson, Maier, & Seligman, 1993; Rodin, 1986). Relatively small but meaningful and important increases in control appear to have a significant impact.

Second, epidemiological research has been conducted in a specific naturally occurring human workplace setting on the effects of what appear to be correlates of uncontrollable stress on human health. These are the Whitehall studies, two extensive series of extended studies on the health of British civil servants (e.g., Bosma et al., 1997; Marmot, Bosma, Hemingway, Brunner, & Stansfeld, 1997). These studies, continued over a considerable period of time, show that lack of control over work circumstances and hierarchy is related to longer term health risk of a variety of kinds. The same studies show that work circumstances that tend to provide more control are related to some better long-term health outcomes.

Biological Foundations of Responses to Uncontrollable Stress

Understanding that exposure to uncontrollable stressors has predictable outcomes on behavior and health of humans and animals made it interesting to develop integrative research approaches to better understand the underlying neurological and biological bases for these responses. As this exploration was occurring, the interdisciplinary field of neuroscience continued to develop, creating a literature on the linkages between brain regions and behavior. This combination led to a research focus on the role that different brain regions play in mediating responses to uncontrollable stress.

A series of experiments in rats, for example, demonstrated the role played in responses to uncontrollable stress by a region in the brain stem called the dorsal raphe nucleus (DRN). The DRN sends projecting neurons to many of the brain regions that are likely to be involved in mediating the observed aftereffects of responses to uncon-trollable stress. The neural pathways involved in transmitting signals from the DRN to these structures use the neurotransmitter serotonin. Following uncontrollable stress, the serotonergic neurons in the relevant regions of the DRN become activated to a far greater extent than occurs in animals following controllable stress.

During uncontrollable stress, the concentration of serotonin in the DRN is increased. This has the effect of sensitizing the serotonergic neurons in the DRN that innervate the target regions in the forebrain and other structures. The consequence of this sen-sitization is that, for a period of days, otherwise mild stimuli or stressors produce large amounts of serotonin, which leads to large changes in the brain regions affected by the DRN. When the serotonin pathways are chemically or surgically blocked, however, then behavioral responses usually related to uncontrollable stress are not observed following uncontrollable stress. Moreover, introducing or stimulating large quantities of serotonin induces such behavioral responses in the absence of uncontrollable stress (Maier et al., 2006).

An implication of these findings is that, following uncontrollable stress, the DRN places limits on an animal's behavior and on its capability to perform work. Because a brain stem structure like the DRN arose at times evolutionarily antecedent to more complex brain structures, such as the prefrontal cortex, the responses to uncontrol-lable stress described above are not likely to be either behaviorally or evolutionarily modern adaptations and, therefore, not limited to humans or mammals.

Although intertidal invertebrates (e.g., mussels) may have defense adaptations, such as stinging, squirting, or chemical responses, much of these animals' environment is, in fact, out of their control. If a predator approaches, the animal cannot run away or pick up a club to defend itself, and it does not have a brain capable of controlling a reaction like running or deciding which way or when to run. Instead, a response to conserve energy following an unavoidable attack may offer the best approach to sur-vival. Lowering activity and locomotion conserves energy and, crucially, reduces the likelihood that its own activities will bring the animal into contact with predators.

Control and the Prefrontal Cortex

Research and logic have led to the conclusion that the DRN region in the brain stem is not capable of serving as a location that processes information about whether a stressor might be controllable, because it does not receive direct sensory input. Nor, due to its very small number of neurons, is it capable of the complex information processing required to decide whether a stressor is controllable or uncontrollable. Rather, the DRN must be part of a more complex circuit: Some other part of the brain must detect uncontrollability/controllability and then regulate the DRN.

The hunt for a brain region that could perform such functions and thus affect the progression of responses to stress has led to regions of the cerebral cortex, responsible for complex information processing. The search more specifically focused on the prefrontal cortex, which is implicated in the "executive function" capabilities of more complex animals, where one would hope there would be a structure that could exert executive control over responses to stressors, perhaps through controlling the DRN region in the brain stem.

Within the rat prefrontal cortex, there is an anatomical region called the *ventral medial prefrontal cortex* (abbreviated as mPFCv), which is involved with behavioral control. Significantly, the DRN brain stem region receives most of its cortical input from this region. The neural connections that reach from the cortical mPFCv region to the brain stem's DRN are excitatory connections, using the neurotransmitter glutamate. Once they reach the DRN these excitatory neurons stimulate interneurons, using the neurotransmitter known as GABA (gamma-aminobutyric acid), to inhibit the crucial serotonergic cells in the DRN that are central to responses to uncontrollable stress. Inhibiting the secretion of serotonin, in turn, has the potential to stop the cascade of responses to uncontrollable stress. The prefrontal cortex thus has the capability to inhibit or control this cascade by stimulating neurons in the DRN that will shut down serotonin release and the behavioral events that follow (Amat et al., 2005; Maier et al., 2006).

Given this cortical capability, does such cortical control actually occur? Experimental results from several labs (summarized by Amat et al. (2005)), indicate that electrical stimulation of the cortical mPFCv region in rats inhibits serotonin units from firing in the DRN brain stem region. This inhibition depends on the neurotransmitter GABA. However, artificially blocking the signals from the cortical mPFCv region from reaching the DRN in animals who are experiencing controllable stress blocks the inhibition of serotonin. These animals behave as if the stress was uncontrollable: They show the behavioral changes normally produced by uncontrollable stress, even in the presence of controllable stress (Amat, Paul, Zarza, Watkins, & Maier, 2006).

Artificially activating the mPFCv produces the opposite effect: Animals actually experiencing uncontrollable stress behave as if they have control, even though they do not have control; they do not experience the effects of uncontrollable stress.

The relationship between the two brain regions, in the brain stem and in the cerebral cortex, also helps explain the sequential relationship in the brain between stress and control. When a stressor is escapable—that is, when the animal has control—activation of the mPFCv causes inhibition of serotonin neurons in the DRN, and the cascade of effects of uncontrollable stress is halted. This sequence provides us with an important distinction: The "default" mode is for the animal to proceed as if an unknown stressor is uncontrollable, unless and until the animal can assess in the cortex that control is indeed present. When control is present, the mPFCv is activated, the DRN is inhibited, and learned helplessness behaviors are blocked.

There are several important implications of these findings:

• The cascade of effects of responses to uncontrollable stress is a powerful adaptation that limits work and movement and conserves energy in the aftermath of an uncontrollable stressor, such as an attack.
• Cortical regions allow animals to biologically operationalize the presence of control: In the presence of control, cortical regions are capable of inhibiting the still extant and powerful consequences of stressors in the brain stem.
• Animals who experience control can then continue to function effectively and perform work in the face of the same stressors that cause profound impairment of activity when these stressors are not controllable.
• Simpler and smaller organisms have fewer options to exert complex control over their environments, especially in the face of attack. More complex organisms can exert behavioral control more frequently and extensively, as they have hands with which to claw and feet with which to run—and a brain to decide in which direction to do so. It is adaptive, then, for a brain region of an extensively motile and more behaviorally and structurally complex organism to be able to also exert control over its response to stressors. The uncontrollable stress response proceeds unless the cortical region shuts it down.
• Two evolutionarily distinct responses to stress thus exist together in the brain, and each can exert its effect on the animal as a whole.

Bidirectional Brain–Immune Responses to Environmental Challenges

We now return to a starting premise of the chapter: Different environmental challenges produce a variety of biological responses in animals—responses that at times have profound (and profoundly different) effects on the capability of animals, including modern humans, to perform work.

Similar environmental challenges can trigger responses in more than one biological system and in more than one way. Different environmental challenges can trigger similar responses—a fact that can have potentially unexpected effects on animals' capabilities to perform work. In this section, we explore responses to the environmental challenge of infection. These responses involve not just the brain but also the immune system, with the two exerting *bidirectional control* over each other.

The presence of an infectious agent within an organism affects the brain and behavior in surprising ways. Equally surprising has been the discovery that this response, evolutionarily adaptive for infection, is also triggered by many acute stressors (Maier & Watkins, 1998, 1999). The response circuit and its effects are therefore part of our everyday lives and work—a fact of which most of us are unaware as we go about our daily lives and also as we think about how work gets done.

Nonspecific Immunity: The "Other" Immune System

The body has, beyond the brain and the nervous system, another extensive system for sensing the environment and responding to it: the immune system. The immune system, unlike the brain and the nervous system, is generally not thought of as being capable of directly affecting behavior; the immune system is often thought of as operating independently of the brain and vice versa. One reason for this supposition of independence is that the blood–brain barrier should prevent large molecules, including chemicals (called cytokines) secreted by the immune system and that have effects on other cells, from reaching the brain.

Most often when we think of the immune system, we think of what is called the *specific immune system*, which provides for the classic antigen–antibody response. The specific immune system, however, has some limitations in responding to infection. An important one is that it takes time—at least three days—for a specific immune response to develop following an initial immune challenge. Something else must initially protect us from infection over that time—it is otherwise unlikely that any of us would be alive or that any of our earliest animal ancestors would have survived long enough to make that possible.

The *nonspecific immune system* is responsible for enabling the animal to begin to hold the line against an invader. This nonspecific immune response starts with the sensing of a non-self entity and the mobilization of cells that can attack non-self cells in that area. Chief among these mobilized cells are *phagocytes*, cells that can engulf foreign entities; primary among these phagocytes is a type of white blood cell called a *macrophage*. Macrophages act directly against the source of threat and also give off signals that help mobilize the body against infection in multiple ways. At the early stages of an infection, it is essential for macrophages not only to engage the invading cells but to signal other cells. Upon activation, for example, macrophages emit nitric oxide gas, which does double duty as a powerful localized toxin and as a signaling molecule. Macrophages also produce and emit more specialized signaling molecules called *cytokines*. "Cytokine" is a coinage from "cyto" for "cell" and "kines" for action or motion. The most prominent and potent cytokine stimulated by pathogens is interleukin-1 (IL-1). Increased levels of IL-1 play a role in signaling increased production and activity of the nonspecific immune system throughout the vicinity of the infection.

Immune-to-Brain Communication

Over time, it has become clear that local mobilization is by no means the only role that IL-1 plays, particularly in the development of an animal's behavioral response to infection. IL-1 also plays a direct role in changed behavior by signaling the brain. There are elevated IL-1 levels in the brain following sickness. Interestingly, injecting IL-1 into the brain of a healthy animal triggers a stereotypical constellation of behavioral responses that occur during sickness, even in the absence of illness. What is not immediately evident, though, is how IL-1, triggered in the periphery of the body in response to an invader, can reach the brain to trigger sickness behavior, because cytokines are too large to cross the blood–brain barrier. Further examination shows that there are IL-1 receptors in the brain and that IL-1 can be secreted within the brain, but that still leaves the mystery of how the signals of infection from the periphery reach the brain to trigger the brain to make IL-1.

One possibility is neural communication. The vagus nerve, which reaches from the brain stem to a wide range of regions in the periphery, sends signals from these regions to the brain stem. One way to test whether this is a pathway utilized is to sever the vagus nerve. When this is done, the sickness response described below is not exhibited in the presence of illness, nor is brain IL-1 elevated (e.g., Watkins et al., 1995). Further research has located receptors for IL-1 on structures surrounding the terminals of the vagus nerve in the periphery and has shown that these receptors carry signals from the periphery to the brain. The responses of the immune system to infection, such as elevated IL-1 levels from macrophage activation, can trigger signal transmission from the periphery to the brain. However, it is important to note that there are also a variety of specialized mechanisms beyond passive diffusion into the brain, such as active transport, that allow some bloodborne IL-1 to signal the brain, so there are multiple signaling mechanisms.

Events in the Brain and Brain-to-Immune Communication

We have so far traced one half of the bidirectional response to infection, from infection site to brain. What about the second half—what happens to the signal once it is received by the brain? The vagus nerve and other methods of communication stimulate several brain stem regions that are the loci for the beginning of a cascade of neural events resulting in profound changes in behavior to the animal and to the animal's capacity to perform work.

The sickness syndrome cascade proceeds from the brain stem to more rostral regions, such as the hippocampus and the hypothalamus, capable of triggering responses which, in turn, are common to both sickness and stress. The first response is *sickness behavior*. The animals' motivation changes and they become less physically active and conserve energy. The second response is a series of physiological adjustments that include fever—an adaptive, protective, and energy-intensive physiological response—which can also be induced by a range of acute stressors that are not infectious agents. The third response is activation of the hypothalamic–pituitary–adrenal (HPA) axis, which mobilizes energy production, making energy available for fever (or, in other circumstances, preparing the animal for an energetically intensive response to a stressor, such as a fight–flight response).

The response also ends up forming a complete loop of communication back to the immune system. Increased IL-1 levels in the brain have several effects in the periphery related to both sickness and stress, including increasing cytokine levels out in the periphery, which completes this circuit. The immune system is capable of triggering a cascade that transmits its responses to the brain. The brain is capable of responding to these changes both with profound effects on behavior and on work and by heightening the level of response of the immune system. In place of a traditional view that behavior is controlled by the brain and communicated to the periphery, it is clear that there is bidirectional communication between the brain and the immune system.

Sickness Behavior and Work in the World

How does an animal respond to the activation of this bidirectional brain–behavior circuit? The following effects have been observed: fever, lassitude, and decreases in social interaction, sexual activity, exploration, and movement, accompanied by decreased intake of food and water (Maier & Watkins, 1998). An effect of sickness behavior is that, taken as a whole, it reduces the capacity of the animal to perform work on the external world. In animals in the wild, this has the beneficial effect of conserving energy and reducing the dangers of foraging for food during a period of weakness. At the same time, due to the coincident activation of the HPA axis, energy is created that can be utilized to allow fever. The shifts that can be produced in energy storage and utilization are, therefore, not unidirectional and crudely overdetermined but are nuanced, powerful, and bidirectional. The coordinated shifts of behavior and physiology that enable energy to be conserved under stress—or alternatively mobilized to fight or flee an attacker—while simultaneously being mobilized for increased macromolecule production and fever production is a remarkable circuit.

This energetic circuit or trade-off provides a basis for understanding the evolutionary linkage we seek between the responses to immune activation and stressors. The combination of sickness syndrome, sickness behavior, and the bidirectional immune–brain system is an adaptation of immense evolutionary value, not only for the protection provided against infection but for the robust adaptation it coveys for dynamically balancing often competing energetic requirements under threat—for example, an external threat would require a fight-or-flight energetic mobilization, but infection requires the opposite. Thus, the bidirectional immune–brain system, once having emerged as an adaptive response to infection, may have then been co-opted over evolutionary time (Gould, 1982; Gould & Vrba, 1982) as responses to noninfective environmental challenges. Because the initial adaptation of the immune response itself is so vital to an animal's survival, its continued presence would make it an ongoing source, once initially co-opted in response to acute stress, for continued and more subtle co-optations adaptive for acute stress—not once but over almost the entire span of vertebrate evolution (e.g., Clack, 2002a, 2002b).

The argument, then, is that the co-option of the sickness response for stress would have occurred for life-threatening fight-or-flight stressors (Maier & Watkins, 1998). Once this co-optation had developed, and these pathways had come to respond adaptively to acute stress absent infectious agents, the range of stressors that trigger these responses could become subtler than fight or flight. Simply exposing an animal to a novel environment, for instance, has been shown to raise cytokine levels and induce fever (LeMay, Vander, & Kluger, 1990). Likewise, some research conducted in this area on acute stress used the same tail shock protocol described in experiments earlier, which itself was found sufficient to induce the effects of illness, including fever lasting for two days (Maier & Watkins, 1998). Nor are rats unique in responding to an acute but noninfective stressor with fever. In humans, fever is produced by a common and work-related behavior: public speaking. And even more common work-place situations are experienced as acute stressors, such as riding in a crowded elevator—on the side of the elevator away from the control buttons (Rodin, Solomon, & Metcalf, 1978).

This leads naturally to the question of what the effects are on an entire population of humans in a work organization in which acute stressors, illness, and sickness are constantly propagating through the population, affecting differing proportions of the populations at differing times and differing magnitudes. It is to these implications that we now turn.

Implications for Work in Animals and in Humans

Several implications result from exploring these mechanisms for responding to stressors in the environment:

1. Stress—like work—is not a biologically unitary phenomenon. Different responses to stressors have evolved at different times. These might respond to different stressors or might respond differently to similar stressors and utilize differing and at times competing pathways and biological systems in their responses.

2. The responses to stressors and environmental challenges explored here have the effect of reducing an animal's capabilities to perform work on its environment and do so in ways that are adaptive to the organism, not necessarily counterproductive in ways that these adaptations have often been viewed.

3. These responses to stressors, the bidirectional brain–immune system in particular, have the capacity to induce vast differences in an animal's or human's ability to work or perform, and these changes can occur for reasons that are not evident either to the individual or to others: Actual illness, exposure to a pathogen, a flu shot, or an acutely stressful event at home or at work can all trigger biologically adaptive performance impairment.

4. In turn, this indicates that stress can and does create substantive *intraindividual differences in performance*; that is, differences in performance within a given individual over a period of hours, days, or weeks.

5. These differences have gone largely unnoted both in organizational life and practice and in research on psychology and on organizations. Psychology in many different subfields has tended to focus on the presence and effect of *inter*-individual differences, or seemingly stable differences between individuals over time. This may have diverted attention from the effects of intraindividual differences.

6. We also note that these differences are by their nature *aperiodic* and *noncyclical* and that many of these differences are triggered by occurrences of challenges in the environment that can be largely *unpredictable*. Just as intraindividual differences may have been obscured by seeming interindividual differences, attention focused on periodic and cyclical changes in performance may have obscured the fundamental nature of aperiodic, noncyclical changes at both the individual and organizational levels, produced by intraindividual differences arising from many sources—responses to uncontrollable stressors, acute stressors, and infection among them.

Responses to stressors and other environmental challenges can thus be viewed in many ways. For the purpose of understanding work in living systems and the implications of stress and stressors on work, it is essential to understand different stressors and environmental challenges as triggering different sets or constellations of combined neurological, behavioral, and physiological responses. Responses to stressors change the work of the organism on the external world. The responses described in this chapter are thus directly linked to the demands for energy conservation and energy mobilization under stress and to the need to integrate these two competing demands within a limited energy supply.

Exploring these processes shows us that the different responses to different stressors produce profound adaptive changes in our physiology and our behavior. These changes produce large and largely unconsidered intraindividual differences in performance that affect individuals and the groups in which they work. Whether a stressor comes from work stress or life stress may make no difference on the effects the stressor has on one's life or work. A single stressor, however, may trigger vastly different responses in different brain regions, distinct biological systems, and physiological and behavioral constellations—responses that will affect each of us as humans and as organisms, in work and in life.

References

Amat, J., Baratta, M. V., Paul, E., Bland, S. T., Watkins, L. R., & Maier, S. F. (2005). Medial prefrontal cortex determines how stressor controllability affects behavior and dorsal raphe nucleus. *Nature Neuroscience, 8*(3), 365–371.

Amat, J., Paul, E., Zarza, C., Watkins, L. R., & Maier, S. F. (2006). Previous experience with behavioral control over stress blocks the behavioral and dorsal raphe nucleus activating effects of later uncontrollable stress: Role of the ventral medial prefrontal cortex. *Journal of Neuroscience, 26*(51), 13264–13272.

Bosma, H., Marmot, M. G., Hemingway, H., Nicholson, A. C., Brunner, E., & Stansfeld, S. A. (1997). Low job control and risk of coronary heart disease in Whitehall II (prospective cohort) study. *British Medical Journal, 314*(7080), 558–565.

Clack, J. A. (2002a). *Gaining ground: the origin and evolution of tetrapods*. Bloomington, Ind.: Indiana University Press.

Clack, J. A. (2002b). Patterns and processes in the early evolution of the tetrapod ear. *Journal of Neurobiology, 53*(2), 251–264.

Gould, S. J. (1982). Darwinism and the expansion of evolutionary theory. *Science, 216*(4544), 380–387.

Gould, S. J., & Vrba, E. S. (1982). Exaptation—A missing term in the science of form. *Paleobiology, 8*(1), 4–15.

Langer, E. J., & Rodin, J. (1976). Effects of choice and enhanced personal responsibility for the aged: A field experiment in an institutional setting. *Journal of Personality and Social Psychology, 34*(2), 191–198.

LeMay, L. G., Vander, A. J., & Kluger, M. J. (1990). The effects of psychological stress on plasma interleukin-6 activity in rats. *Physiology & Behavior, 47*, 957–961.

Maier, S. F., Amat, J., Baratta, M. V., Paul, E., & Watkins, L. R. (2006). Behavioral control, the medial prefrontal cortex, and resilience. *Dialogues in Clinical Neuroscience, 8*(4), 353–374.

Maier, S. F., & Seligman, M. E. P. (1976). Learned helplessness: theory and evidence. *Journal of Experimental Psychology. General, 105*, 3–46.

Maier, S. F., & Watkins, L. R. (1998). Cytokines for psychologists: Implications of bidirectional immune-to-brain communication for understanding behavior, mood, and cognition. *Psychological Review, 105*, 83–107.

Maier, S. F., & Watkins, L. R. (1999). Bidirectional communication between the brain and the immune system: Implications for behaviour. *Animal Behaviour, 57*, 741–751.

Marmot, M. G., Bosma, H., Hemingway, H., Brunner, E., & Stansfeld, S. (1997). Contribution of job control and other risk factors to social variations in coronary heart disease incidence. *Lancet, 350*, 235–239.

Peterson, C., Maier, S. F., & Seligman, M. E. P. (1993). *Learned helplessness: A theory for the age of personal control*. New York: Oxford University Press.

Rodin, J. (1986). Aging and health: Effects of the sense of control. *Science, 233*(4770), 1271–1276.

Rodin, J., Solomon, S. K., & Metcalf, J. (1978). Role of control in mediating perceptions of density. *Journal of Personality and Social Psychology*, *36*(9), 988–999.

Seligman, M. E. P., & Maier, S. F. (1967). Failure to escape traumatic shock. *Journal of Experimental Psychology*, *74*(1), 1–9.

Watkins, L. R., Goehler, L. E., Relton, J. K., Tartaglia, N., Silbert, L., Martin, D., & Maier, S.F. (1995). Blockade of interleukin-1 induced hyperthermia by subdiaphragmatic vagotomy: evidence for vagal mediation of immune-brain communication. *Neuroscience Letters*, *183*(1–2), 27–31.

Reflections: On Exploring Work in Living Systems

After a long exploration and a long journey, how might one begin to sum up and open up what one has seen? Part of what explorers endeavor to do is to open up new country. We hope that our contributed chapters have given you that same sense of exploration that we experienced in developing this volume. In this chapter, we hope to open up new country for further exploration.

If one returned from a scientific exploration, there would be several ways to describe what one had seen to open up country. One could describe broad *themes* that emerged over the course of the scientific exploration. One could draw together cogent *observations* from the exploration that help make sense of what was seen. One could *visually survey and contrast* a sample of the phenomena encountered. Finally, one could develop a perspective that *moved between levels* of organization in the spirit of integrative research.

The views we have chosen to sum up this exploration are labeled: Perspectives on Exploring, Observations, Design Interlude, Work Integrates Life, and Life Integrates Work. The material in these sections has been generated by looking back over the experience of developing this volume, of exploring this subject as a group and as individuals, by looking back over the country we have covered and by recollecting the time we have spent working together. This material is, therefore, the product of the editors and contributors collectively. All of the editors contributed to each of the sections that follow, and we have drawn from the contributors' materials in creating all of these portions. Christina De La Rocha coordinated the Observations section, Alan Blackwell the Design Interlude section, Simon Laughlin the Work Integrates Life section, and Robert Levin the Life Integrates Work section.

What follows is decidedly and deliberately exploratory in nature. We may, for example, bring research from two contributors together in ways that might not have been visible from the vantage points of the individual chapters. Of course, we might be doing so correctly or incorrectly. For this reason, we want to state clearly that although we have drawn from the contributors' materials in creating the Reflections material, the accuracy of any material here is the responsibility of the editors, not of the individual contributors. For presentations of content, for explanations, and for references, the reader is referred back to the relevant portions of chapters 1–9. References from the chapters are not repeated here. Lastly, we have not attempted here to summarize the book; instead we start with some of the topics and conclusions of the previous chapters and use them to explore further.

Having indicated the nature of the exploration that lies ahead and, we hope, the openings, and having delivered our stern warning of how our account of what lies ahead was formed and why, we want you to venture forth on a friendlier note: Onward. And enjoy.

Perspectives on Exploring

We start by presenting a few of the many perspectives that emerged over the course of our exploration. Here are three perspectives we gained along the way that we did not have at the outset and that we wish to present to you.

Trade-offs Matter

Our original plan for exploring the study of work in an integrative way placed no particular emphasis on trade-offs. Throughout the development of the book, however, it became clear that trade-offs matter a great deal to how work gets done. We may not see or understand all of the trade-offs that matter, but they will be acting as powerful forces on how work gets done because they tend to be inherent and inescapable.

Consider, for example, two of the chapters that explicitly dealt with trade-offs as a part of their research. Energy–information trade-offs, we now understand, constrain how much information can be processed in a sensory system, in a brain, or in an organism. Because the work conducted by a living system is generally energy directed by information, this means that energetic constraints on information will place limits on the work that can be done by an organism. The work that is limited is not only physical work. Energy–information trade-offs show us that working with information is not exempt from limitations imposed by available energy. Humans' increased capacity to offload information externally certainly increases the capacity to store and process information externally, but the capacity of a brain to work with information is still energetically constrained.

Now consider the trade-offs that have been described between performance and yield. Performance–yield trade-offs emerge in the nature of any system that gets made—whether it is manufactured or the offspring of a living system. The work performed by an organism is, therefore, subject to performance–yield trade-offs.

Performance–yield trade-offs tell us that there is no form of egg from which 100% of chicks will hatch, nor a form of fox that will catch 100% of chicks. (The question of whether there is a form of coyote that will ever catch the roadrunner is for a different forum.)

The theme of trade-offs existed in less explicit ways in other explorations as well. The presentation of responses to stressors shows that these responses often take the form of trade-offs between conserving energy to combat an environmental challenge, such as infection, or mobilizing energy to respond to a stressor, such as a threat of attack.

Performance Envelopes

Our second theme is related to trade-offs: It is what we have called *performance envelopes*. The general notion of a performance envelope is familiar from reading about test pilots—they test and push the absolute performance envelopes of an airplane under high-demand conditions, and sometimes they push the airplane past one of its absolute limits.

The concept we wish to raise to open up the integrative study of work is that the boundaries of or even the existence of a given performance envelope are not always obvious. One of us became aware of such an envelope flying high above the Atlantic, reading in a magazine, as it happened, about one of the hazards of early transoceanic flight. It turns out that, at high altitudes and high speeds, it is possible for an airplane, by climbing into thinner air where it can accelerate continually faster and climb higher into even thinner air, to reach a speed and altitude beyond which it cannot go any faster and cannot go any slower without stalling. Because neither the individual speed nor the individual altitude parameters are violated along the way, this effect, and the performance envelope that created it, were apparently learned about by experience. So, too, with much of life and with much of work.

We have also seen that such performance envelopes operate at many levels of work, and we suggest that more awareness of these envelopes would provide fruitful areas for research as well as for application. The process of chemiosmosis creates a fundamental performance envelope on all biological work. Because chemiosmosis is always grinding forward to maintain ATP concentration at high relative concentrations in the cell, which is essential, in turn, to any work being done by the cell or the organism, the capacities of the chemiosmotic system and its unstoppable requirements for energy will define boundaries of many performance envelopes. For example, energy limits muscular work and the ability to use information to direct work. Inherent performance–yield trade-offs impose another similar constraint on the absolute performance of any organism. In human design, attention investment represents a performance envelope resulting from a moment-to-moment trade-off, one that relates to energy–information trade-offs.

Work Integrates

The previous two themes are examples that establish one of the starting premises for the book: that it would be fruitful to explore phenomena related to work across disciplinary lines, across levels of organization, and across species. We believe that the primary chapters have already well established the fruitfulness of this premise.

Here, though, we want to note a somewhat different effect of how work integrates. The preceding two themes have emerged from studies of particular organisms—and of nonliving systems. These themes can then be the source of research and study in many other organisms and in many natural and human settings, as well as the source for integration with the needs for development of solutions to practical issues in diverse areas. The integration in the two preceding themes and in the primary chapters causes us to return to a theme that was quite the opposite of how the book began but emerged through both the process and the content. We believe strongly that it is imperative to move away from a hierarchical or progressive approach that views human work as distinct from that of other organisms and allow both research and the development of problem solutions to benefit from the full range of research into living systems. When we better see the commonalities in how work gets done across widely varied forms of living systems, we also gain a better understanding of the particularities of a given species.

A theme emerging from our exploration, then, is to call for continued research on how work gets done in a wide variety of living systems, not only in humans and not only in our apparent "nearest neighbors." In making this call, the importance of a rigorous approach to integrating research across living systems must accompany it. We do not seek holistic solutions beyond what can be learned from living systems themselves or for such research to become a basis for argument by analogy. Instead, it is essential to look carefully for possible linkages and then test them by careful investigation. But we feel strongly that an integrative approach can open up new vistas in many fields of inquiry and feel just as strongly that it is essential to solving many problems in human work and in the never ending trade-off, sometimes called sustainability, between our own sustenance and maintenance and that of the rest of the natural world—or perhaps better, the working world—on which our lives and work depend.

One way to search for such links is to systematically search for recurring patterns in the way that work is done by living systems.

Observations

There is both an absolute limit and an operational limit to the amount of work that can be done in any system, living or nonliving. The absolute limit is set by physics. The laws of thermodynamics, which govern the conversion of energy from one form to another and the conversion of energy into work, define how much energy can be extracted from a given material and the efficiency with which that energy can be converted into work. Operationally, however, the amount of work that gets done in a system is smaller than the thermodynamic limits, being constrained by infrastructural or supply limitations and trade-offs between speed and energy efficiency or cost and reliability.

Although there are instances in which life converts energy into work close to the thermodynamic edge, overall, living systems employ processes feasible and functional rather than maximal. The molecular mechanisms of aerobic photosynthesis are only 3% to 5% efficient at converting solar energy into sugars that can be used to fuel cellular work. And a person walking a mile on a city street burns through approximately three times more energy than he or she would have pedaling a bicycle, which would also have gotten him or her to the destination faster to boot.

Because thermodynamics is not the only force governing work in living systems, and because living systems do work on many different levels of organization, it is worth considering similarities in the problems, pressures, trade-offs, and working strategies employed by living systems across various levels of operation and scales of time. Can we make observations that characterize work as done by living systems? Can we use knowledge about the trade-offs, limitations, and work strategies deployed on one level in living systems to understand the patterns of work emerging at high levels of organization?

During our explorations of work in living systems as we crafted this book, we noticed several recurring patterns, or observations, and describe some below.

Patterns of Work of Living Systems

Observation 1: Work requires energy.   Although the consumption of energy for work absolutely underpins the work of living systems, few of us working outside of the biological sciences know how life transforms energy into work. To give a brief summary, living creatures gather energy from the environment and store it until use, when it is fed into metabolic machinery in a form like the sugar glucose. This organic matter is then oxidized to yield carbon dioxide (CO_2) and energy that will be harnessed for work. The oxidation of one mole of glucose, weighing 180 grams (g) (and about the amount of glucose in a donut, as it happens), gives off a fair amount of energy, 686 kilocalories (kcal): To get a feel for this amount of energy, take seven jiggers of rum and set them on fire.

Within living creatures, however, organic matter is not converted into CO_2 in one fell energy-releasing swoop, but, rather, it is oxidized in a series of smaller, lower energy steps. The smaller packet of energy released at each step is used to do the work necessary to create zones of a high concentration of hydrogen ions (protons, H^+) within the cells of eucaryotes and on the outside of the membranes of archaea and bacteria. These protons then diffuse out of the areas of high concentration, and the energy of their passage is used, in chemiosmosis, to construct adenosine triphosphate (ATP). ATP, an energetic, unstable molecule that delivers energy by binding directly to enzymes, proteins, and chemical reactants, is the energy source proximally responsible for work in living systems.

Observation 2. The conversion of energy into work is never 100% efficient. Thermodynamics requires that whenever work is done, some of the energy put in must go toward increasing the entropy, or disorder, of the universe. The amount of heat given off by a running engine or a person exercising are examples of the energy released during the process of work that is not getting work done. Likewise, 16 of the 686 kcal released by oxidizing 180 g of glucose are converted into heat, leaving 670 kcal, or 98% of the released energy, to fuel work.

According to thermodynamics, when plants put CO_2 and water together to make 180 g of glucose, the minimum amount of energy they must invest is 686 kcal. But it would be a happy plant indeed that was that proficient at making glucose via photosynthesis. Because the products of photosynthesis (glucose and oxygen) have a higher free energy associated with them than the reactants (CO_2 and water), this chemical reaction constitutes work and, under ideal conditions, takes 2,880 kcal of energy, in the form of 48 moles of photons, to produce 180 g of glucose. Thus, plants, even when operating under ideal conditions, could only be 24% efficient at converting energy (photons) into work (the construction of glucose). It should be noted that ideal conditions are rather different from normal operating conditions, where energy losses occur at almost every step in the process. Operationally, photosynthesis tends to be 3% to 5% efficient at converting sunlight into sugars.

Observation 3: Life directs energy toward the accomplishment of specific work. This energy may be used to overcome either thermodynamic barriers to work (i.e., $\Delta G>0$), kinetic barriers to work (i.e., slow reaction times), or both. It is easy to overlook the fact that work is not a unique feature of life but rather the simple consequence of there being energy at hand. Without any biological intervention, mountains are raised over geologic time, fuelled by heat from the Earth's interior. Likewise, wind, weather, and gravity are by themselves sufficient to raze those mountains to the ground.

While not all work requires life to direct its execution, life often does direct the execution of work. This habit of directing work is one significant trait that distinguishes life from nonlife. When biological systems engage in work, energy does not just drive whatever nonspontaneous reactions are at hand, but is directed toward the accomplishment of specific tasks. These tasks may be carried out at a faster rate than they would occur naturally, or they may be work that would not otherwise occur. We, as human beings, have designed and built geothermal power plants to collect the same heat energy that drives mountains up into the sky. We convert this energy into electricity, which we, in turn, use to do many things—from run computers to construct skyscrapers. In constructing skyscrapers, we are lifting rocks (like concrete and gypsum) up from the ground millions of years faster than even the fiercest continental collision could do in terms of mountain building.

Observation 4: Life gathers energy and uses it to do work in order to survive, thrive, and reproduce. Although it is obvious that some of the outcomes of work include promoting the survival, well being, and propagation of oneself and of kindred, it is easy to lose sight of this in the daily grind of going to the office to feed our yearnings for fashion, gadgets, travel, and cars. For most organisms, however, work is almost wholly devoted to survival and reproduction (e.g., maintaining cellular functions, gathering food, building dens or nests, migrating with favorable conditions, gestating and rearing offspring). For humans, it was until quite recently true in any part of the world that some failure to accomplish work (e.g., produce sufficient harvest due to war or bad weather) meant famine, loss of reproductive capacity, or death. In many parts of the world, this direct link between work and survival still exists. As human numbers continue to increase, it may yet be true for the entire population again.

Observation 5: Work may be more than just mechanical in nature. The cells of all living organisms carry out three different types of work: work that is mechanical in nature, chemical in nature, or that maintains gradients. Mechanical work involves things like the beating of flagella for locomotion or the dragging of duplicated chromosomes to opposite ends of a cell during division. Chemical work involves driving reactions that are not energetically or kinetically favorable, such as the construction of large macromolecules, like enzymes and DNA, critical to the functioning of the cell and organism. The third type of work is that of pumps that maintain chemical and electrical gradients across membranes. This work makes chemiosmosis possible, allowing cells to accumulate necessary nutrients even when the nutrients exist at dilute concentrations in the external environment. This work also, by restoring the electrical potentials of neurons following the transmission of a signal, allows the nervous system to process information.

These three types of work are proximally fuelled by the breakdown of ATP. Mechanical work and the working of pumps inside cells occur when ATP binds to a protein that catalyzes the breakdown of ATP and release of energy. This causes the protein to change shape, and the change in shape causes the desired work to be done. With flagella and muscles, the binding of ATP causes protein "arms" to bind and release and bind again to structural fibers, sliding them past one another, bending a flagellum or contracting a muscle. Ion pumps, which span from one side of a membrane to another, also change conformation when bound to ATP. When an ion pump changes conformation, it pulls an ion or two from one side of the membrane to the other. When chemical reactions are driven by ATP, it is the released phosphate that binds to the reactants, bringing energy with it. Thus energized, the reactant is able to go forward through a reaction that could not have otherwise proceeded or would have otherwise happened extremely slowly.

Observation 6: Work in living systems requires infrastructure, supplies, and equipment. Another thing that distinguishes life from nonlife is that living organisms invest energy into producing infrastructure, supplies, and equipment with which to carry out specific work or to accomplish work more effectively. As described in Observation 5, the interaction of energy (in the form of ATP) with the structure of a muscle provides an organism with the leverage to move its skeleton. Phytoplankton must have, in their cell membranes, uptake systems specific for the acquisition of nitrate and phosphate if they are going to obtain these basic building blocks of proteins, enzymes, and DNA. All cells must also have the protein complexes of the electron transport chain embedded in strict association in a membrane if they are going to take energy from respiration and, via chemiosmosis, turn it into ATP.

Observation 7: There are physical (chemical and infrastructural) limitations to the quantity and type of work that can be accomplished from a given parcel of energy in a given amount of time under a given set of conditions. Exactly as there is an upper boundary to the speed and distance that people can run due to the structure of their bones, the amount of muscle they have, how much fat and carbohydrate they have stored up for the creation of ATP, and how efficiently their body is able to whisk away the lactic acid that builds up in their muscles, there are chemical and infrastructural limits to all work that gets done in living systems. The maximum rate at which a eucaryotic cell with a flagellum can propel itself depends on factors like the rate at which the critical enzyme protein can acquire ATP, hydrolyze it, and release ADP, the rate at which that protein can make or break bonds with the microtubules it sends sliding along each other, and the distance it is able to slide the microtubules for each ATP hydrolyzed. Also important is the amount of force that the whipping flagellum can generate and the efficiency with which that whipping force is translated into forward motion; a flagellum that is inadequately placed or controlled will send a cell spinning only in circles.

Infrastructure can also limit the efficiency of the building up of chemical and electrical gradients. If, for example, mitochondrial membranes are leaky with respect to the high concentrations of protons set up within them during chemiosmosis, more energy must be invested to maintain the steep gradients necessary for the cell to regenerate ATP. Likewise, the number of protons that can be pumped across the membrane against a concentration gradient, for a given amount of energy, is tied to the molecular composition and three-dimensional structure of the protein complexes that serve as the pumps. A small change in either could have a significant impact on the efficiency with which food energy is converted into proton-motive force and, ultimately, ATP.

Observation 8: Information may be used to increase the efficiency of work. The collection and utilization of information is another hallmark of work done by living creatures. This is nicely exemplified by brains, which have evolved to direct the work of animals (including humans). A brain is capable of collecting and storing information, learning, making decisions, driving muscles, and controlling other physical responses to the environment. A brain also directs work that takes place within an animal, especially that which requires the coordinated, and, at times, bidirectional, action of systems like the immune system, such as the response to infection.

Observation 9: Innovations can be developed to increase the output of work per unit time or to broaden the conditions under which work may be carried out. The last 4 billion years have witnessed the continual development of the tools and techniques with which living systems get work done. The accumulation of innovations over this span of time has resulted in a massive increase in the volume of inhabited ecosystems, in the total mass of living creatures, in the number of different ways they get work done, in the total amount of work being carried out in living systems, and in the amount of energy fluxing through the biosphere since the first dawn of life on Earth.

One of the innovations that had a huge impact on the size of the biosphere and ways in which living systems do work was oxygenic photosynthesis, which oxygenated the atmosphere, making aerobic respiration possible. Because aerobic respiration releases roughly 20 times more energy from the breakdown of glucose than do the various types of anaerobic respiration, the biosphere was able to fuel a lot more work from the organic matter within it that was available for consumption. This increase in the efficiency of respiration set the stage for the evolution of complex, multicellular forms of life with energetic lifestyles. It also paved the way for the development of the vast array of body types and morphologies of animal life that are, in themselves, innovations for doing many different types of work effectively.

Observation 10: Because innovations are both sequential and cumulative over time, the infrastructure that has been developed in the past sets boundaries on the innovations that can be developed in the future. This is especially true of living systems for two reasons. The first is that the basic, critical biochemical processes of cells (e.g., photosynthesis and chemiosmosis) rely on tightly interlocked sets, series, or cycles of reactions and enzymes that were developed billions of years ago in organisms belonging to species that no longer exist and under environmental conditions considerably different from those prevalent today. Many enzymes, for example, have as reaction centers metals that occurred in high concentrations in the anoxic waters of the ocean of the early Earth but are rare in today's oxygenated seawater. The lack of these metals now often limits the growth of phytoplankton and bacteria. Although it would seem that enzymes that utilize other, more widely available metals could develop, it is hard to see how they could arise and evolve in the shadow of the poorly functioning but nonetheless vitally important extant enzymes. Instead, organisms have evolved metal-binding ligands that they release into the environment and, that, in part, serve to lengthen the residence time of the rare metals in seawater.

The second reason that innovations are so frequently constrained by previous innovations in living systems is that, during evolution, structures and processes initially developed for one task often become co-opted into other processes and used to carry out different tasks. How different in structure would bird wings be if they did not have their proximal origins as the forelimbs of dinosaurs? Likewise, would stress and sickness provoke similar sickness behavior if interleukins, key elements in the immune system in mammals and invertebrates, had not been co-opted from the immune response by the stress response (or vice versa)?

Observation 11: Work has consequences.   Work has consequences, not all of them intended. Although some unintended consequences are positive (e.g., the temperament of local environments through niche construction), more often these consequences are undesirable. It is unavoidable, for example, that work produces waste, most if not all of which is often toxic to the organism producing it.

Although we have not yet transformed the Earth to the extent that oxygenic photosynthesizers did, it is fair to say that human work has had widespread consequences. On the one hand, our agricultural work supports the lives of billions, and we have created great cities, art, culture, and music. But our work is also driving deforestation, acid rain, ozone depletion, global warming, and the extinction of a considerable number of plant and animal species. The thought that human work can have drastic and far-reaching negative consequences is nonetheless rejected by a somewhat surprising number of people. But the conversion of energy into work never comes without thermodynamic, chemical, and physical consequences. It is worth accepting that our work does, for better and worse, change the world and that, where those changes are unacceptably detrimental, we must either work less or put effort, energy, and expense into diminishing the impact of working.

These observations of recurring patterns in the accomplishment of work by living systems have been offered for your consideration—not as laws or rules, but as suggestions of sights to see as you tour the world of work outlined in this book. As you read this book, and move on to explore more widely, we hope that you will consider whether these observations help to represent ways that living systems operate. We also hope that these observations will help to highlight the importance of physical, biochemical, and cellular operations on the work that is done at the organismal and social level of biological systems. The observations are likewise one foundation for the development of an integrative view of work.

Design Interlude

The previous section has described work as a phenomenon in living systems. One might now ask whether there is an opportunity for action: Is it possible to intervene in organizations, invent new technologies, or refine our future understanding based on the analyses and content in this volume? Using new knowledge as a starting point for intervention might be an opportunity for design. However, the role of a professional designer is often a transient one: refining concepts, opening discussions, or synthesizing competing demands and constraints. This interlude can, therefore, be seen as a design intervention, albeit of a particular kind—contributing to the design of this volume and of the reader's experience. The result is an invitation to the reader to pursue possible directions arising from a design research perspective.

The term *design research* has several meanings. It can mean the study of designers, as in chapter 6. It can mean the research required during the process of designing, as was required to prepare illustrations for chapters in this volume. A third meaning of design research applies to this interlude. The act of designing is itself a kind of research, in which new ideas arise by exploration through making. In both academic art and design, this practice-based research has gained significance to defend disciplines where knowledge is tacit or nonverbal and is not easy to publish in conventional form.

This main content of this design interlude is a piece of design practice—the drawing of figure R.1. The early stages of its creation followed those discussed in chapter 6—understanding the context and requirements, collecting material, and creating an original visual concept via a series of pencil sketches. The final form is perhaps unusual for scientific illustration. The intent of this diagram is to reflect on the themes of the book, to present the results of reflection, and to do so in a manner that opens up discussion, rather than closing it down by presenting a definitive structure. In written and spoken language, ambiguity and inconclusiveness are commonplace and, to some degree, appropriate and intended in a reflection such as this. Ambiguity in visual presentations, especially diagrams, is less expected, so the fact that it is intentionally ambiguous requires some explanation.

The main feature of the figure is a number of paired concepts in the study of work, categorized into two kinds of pairings. As noted above, work often involves paired trade-offs. However, the designer also noticed that the classical definition of physical work—conversion between forms of energy—could be interpreted more broadly to include conversions between activities, substances, and artifacts. These *transformations* are contrasted with *trade-offs* in figure R.1, while also implying visually that transformations and trade-offs might be similar kinds of relationships. The two relationships are generally organized to either side of the page but are also scattered across the page, requiring that a reader not simply read off the conclusions of the figure but inspect and explore it to identify which pair is of which kind.

The background to the figure is intentionally unconventional and "unscientific." It indicates that the plane of the page might be interpreted as a metrical space—one that has two dimensions, but intentionally orienting these dimensions at inexact angles and using huge arrows rather than regular Cartesian axes. These are shaded to indicate that their direction and end points are indeterminate. They refer to the varying physical scales at which different paired phenomena can be observed and the varying cycles over which conversions are completed. It seems that larger scales are quite closely correlated with larger cycles in the figure, but this is for the reader to reflect on. The arrangement of the pairs is not intended to be definitive and is presented in a casual, scattered way in order to invite inquiry.

This explanation of the process and the product draws attention to the visual rhetoric in scientific and technical illustration and representation. There are chapters in this book in which the power of a visual image has been at the center of a scientific position or a change in thinking. The image here is a different contribution, but the design interlude offers both a change of pace and a prompt to those who wish to explore new perspectives, a graphic elicitation inviting readers to explore and develop their own conceptions of work in living systems.

Work Integrates Life

We are immersed in work: the pattering rain, the hiss of windblown sand, water rippling rocks smooth, the glowing red of the vivid sunset painted by scattering ash—the distant legacy of volcanic eruption. All becomes quiet; we close our eyes, we feel at rest, yet still the ground moves beneath our feet, conveyed by the convection of molten rock. There is no rest—energy flows back and forth, gently and violently,

---

Figure R.1 (facing page)
A synoptic view of some of the themes that have been presented throughout this volume, represented in terms of paired entities, either transformations or trade-offs, that are inherent in work. Both transformations and trade-offs occur at a range of physical scales and in cycles over a range of temporal intervals. The organization of this figure is intended to invite reflection and/ or debate with regard to opportunities in the future study of work. Original illustration by Alan Blackwell.

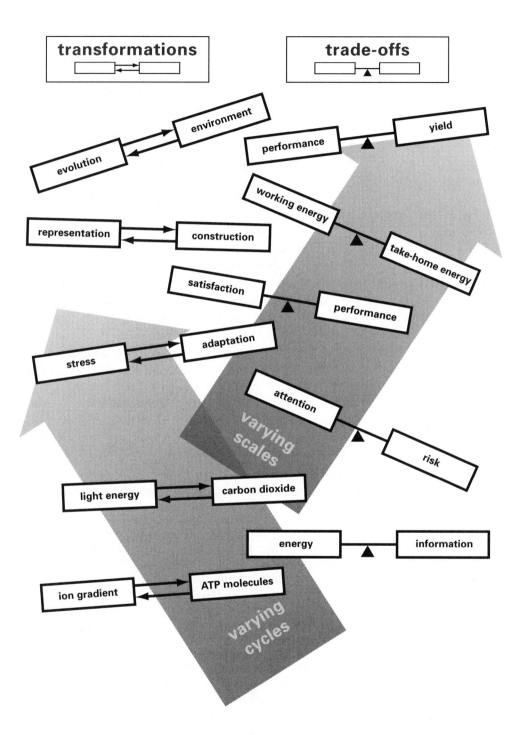

smoothly and abruptly, hotly, heavily, lightly, drily, wetly, steamingly, shockingly, soundly, sulfurously, all driven by the inequalities of an uneven planet, rocking around in a changing universe.

Within this inanimate flux dwells order: living systems that work to reproduce and evolve as conditions change. Reproduction and evolution tie together many of the forms of work in living systems. This integration takes three forms. Processes operating at different scales, from the microscopic—atoms and molecules—to the macroscopic— organisms, groups, societies, and ecosystems—are linked together. Second, the competition for finite resources promotes the efficient organization of these systems. Third, common descent endows living organisms with sets of similar mechanisms. We expand on each of these in turn.

The first form of integration—across multiple scales—reflects the organization of matter, from particle to atom to molecule and so on. In biology, it is integration across scales that makes life tick. Genes root biological systems at the atomic and molecular levels, and organisms must of necessity build upward. The molecular structure of the DNA or RNA in a genome defines the individual and its species and enables the individual to self-organize into the functioning self. Genes act through the agency of the molecules that they specify, such as transcriptions factors and proteins. Again, these agents are molecules, and they self-organize to construct the larger scale systems, such as branches, limbs, and gonads that enable reproduction. And, so, back down to genes.

Animals' prime mover, muscle, provides an example of the integration of molecules into working systems of ascending scale and power. Force is generated at the molecular level by the flexible leg of the motor protein molecule myosin, stepping along the scaffolding protein actin. Stepping is a cyclical process. Each tiny step, of approximately 7 millionths of a millimeter, generates a force of a few piconewtons using the energy gained from the hydrolysis of an ATP molecule. To produce useful forces by useful amounts, billions of myosin and actin molecules are arranged to work together, in muscles.

A muscle's structure is hierarchical. The basic contractile units, myosin molecules and actin molecules, form into myosin filaments and actin filaments. Several hundred of these filaments are then arranged in an interdigitating array so that each myosin foot has an actin to step along. This contractile building block, the sarcomere, is about two thousandths of a millimeter long and one thousandth of a millimeter across, and its structure enables thousands of myosin feet to produce a larger force by marching together, in ranks. Sarcomeres are then lined up in single file to form a fibril that stretches from one end of the muscle to the other. This arrangement amplifies the contraction. For example, if each sarcomere contracts by one half of a thousandth of a millimeter, then 100,000 sarcomeres, placed end to end to make a 25-cm-long muscle fibril, can contract by 5 cm. To amplify the force of contraction and ensure that all myosins work together, thousands of muscle fibrils are packed into a muscle fiber that contracts as a whole. Then, thousands of these larger muscle fibers are arranged in parallel to form a single muscle, such as your calf. This hierarchical structure of units of ever-increasing size harnesses the contractile forces generated by billions of molecular feet to produce a force that can lift your body.

Thus, a muscle exemplifies the way in which living systems use molecular mechanisms (specified by genes) integrated across scales to achieve macroscopic effects and can thereby build large structures that do work on the animal and its surroundings. By constructing nests, other dwellings, tools, and machines, individuals and social groups have extended the scope of the work performed by living systems to niche construction and technology.

The second type of integration involves the organization and coordination that enables systems to perform their roles efficiently. The rationale for efficiency in biology is straightforward. The role played by a biological system, such as a brain, a limb, or a chloroplast, contributes to the fitness of the parent organism, where fitness is defined as lifetime reproductive success. During evolution, natural selection favors systems that promote fitness. Because many of the systems are using limited resources to perform their roles, such as materials, time, and energy, systems that are more efficient, in the sense that they do more for the same cost or the same for less cost, will be selected. The continual selection of variants that work slightly better produces designs in nature that resemble the purposeful, premeditated efforts of engineers but do not involve a sense of purpose and forethought.

Many of the elements of design have been observed in this volume. We see physical constraints that create choke points in systems, such as the weak affinity and low turnover rate of the carbon-fixing enzyme in photosynthesis, Rubisco. This constraint has proved difficult to overcome, but, during the course of evolution, many other limitations have been swept aside by innovation. For example, a biophysical constraint to the speed with which action potentials travel along a neuron's axon was eased by the development of a custom-made insulator, the myelin sheath. But the design envelope is bounded by a second set of constraints: constraints on the ability to innovate. There are limits to the range of materials that cells can produce and to the forms of structures that cells can build. There is no question that placing copper wires in neurons would improve their energy efficiency by making them better conductors of electrical signals, but it is difficult to conceive of a metabolic pathway for manufacturing high-purity metallic copper at room temperature. Turning to form, for developmental reasons that are a legacy of our evolutionary history, our retina, like that of all known vertebrates, is built back to front. Light has to pass through a layer of neurons to get to the photoreceptors, and this blurs the retinal image. A reversed retina would perform its work better, but several hundred million years of vertebrate evolution have, as far as we know, failed to correct this glaring design fault.

Constraints are also eased by devoting more resources to the execution of rate-limiting steps, such as increasing the number of ion channels to reduce noise limitations in neurons. The strategy of increased resource allocation generally involves trade-offs between increases in cost and increases in performance. Quantitative relationships between cost and performance can, in principle, be exploited to make trade-offs that improve the ratio between benefit and cost. Engineers use techniques such as design centering to find the optimum balance between cost and benefit, and it is not unknown for biological systems to reach similar optima. The application of design centering to chip manufacture revolves around the confinement of error within an acceptable range, and this leads to the use of information to reduce uncertainty. Information is also used to adapt the response of the system to changes in operating conditions so that it continues to operate effectively and efficiency. Here, again, there are trade-offs between the costs of processing information and the benefits that accrue from being better informed. We humans devote 20% of our basal metabolic rate to powering our brain, and this evolutionary investment in information mirrors the increasing reliance of our societies and economies on rapid access to large amounts of information.

This volume was originally intended to look at work across many organizational levels. We would highlight the common organizational principles of scale and efficiency by examining a range of organisms, from *E coli* to *H sapiens*. We have extended somewhat further than we originally had thought. The introduction of the energetic process of chemiosmosis and its incorporation at the root of the Big Tree of Life produced the understanding that work shares much in common across the entire range of living systems. As a result of our explorations, we recognize that similar mechanisms exist across all three domains of life, Archaea, Bacteria, and Eucarya. Thus, all biological work has its roots in a relatively small set of mechanisms, inherited through descent from a common ancestor.

But, you will ask, how does an integrative approach produce a better understanding of work? The short answer is that it encourages us to innovate in how we ask questions of living systems about how work gets done, to make the new connections and fresh interpretations that lead us in new directions. The next section raises some of these possibilities by looking across the chapters in this volume.

Life Integrates Work

In this section, we integrate many of the factors and principles about work in humans and in other living systems that we have learned about through the chapters in this book, with two plausibly real-life scenarios: the improvement of a chip factory and the design and operation of a neurosurgery wing. Our purpose here is to see in these examples how the integrative pictures of different aspects of work, so thoroughly developed by the contributors in each of the chapters, together interweave and produce a far richer and more complex understanding of our two scenarios then might have been possible before beginning the integrative exploration of work in living systems.

Consider, as a starting point, the topics introduced by Lightner, who shows how techniques and principles initially developed for chip manufacturing can be applied to living systems, such as a hatching chick and the Pony Express. Doing so captures one direction of the power of an integrative approach to work in living systems: These approaches were initially developed for human manufacturing systems, and we learn that we can integrate them with research on living systems to better understand how living systems perform work.

Can we also move in the other direction? Can we take principles derived from research on organisms and thoughtfully and validly integrate this research with problems in human-designed work? We think so and will develop just such an example here, first applying the energy–information trade-offs identified by Laughlin in work on flies, and integrating the principles of these trade-offs with Lightner's work from chip manufacture to develop approaches for improving that manufacturing process. Once we have done so, we will move to a different but analogous setting—the design and operation of a new hospital neurosurgery wing and there add principles developed by other contributors that might apply to different aspects of the work involved. In this way, we can provide an example of how findings from the integrative study of work can be integrated back and forth across domains.

We make the first step on our journey, then, by integrating the principle of energy–information trade-offs described by Laughlin with the concepts of performance–yield trade-offs and parameter variation described by Lightner. The principles of performance–yield trade-offs and of design centering indicate that it will be effective under a given set of conditions for the performance parameters of a system, such as a chip manufacture foundry, to be backed off of the efficient frontier by an amount equivalent to the magnitude of the parameter variation. If parameter variation were to be zero, which is impossible, then it would be possible for the chip foundry to be designed or operated on the efficient frontier, rather than some "distance" behind it. Because the magnitude of the parameter variation represents that distance, it follows that that distance represents the performance, which might be measured in units produced or in money earned, which is traded off in order to obtain acceptable yield—to obtain reliably acceptable yield.

This means that, in the event the parameter variation were to be reduced, the performance parameters could move closer to the efficient frontier. The foundry, for instance, could produce more units per day or with lowered costs or increased profits. What investments might make it possible to reduce parameter variation? The energy–information trade-off in Laughlin's chapter indicates that parameter variation can be reduced through investments in *information*. Information, in Laughlin's treatment, directs energy to do work. In the chip foundry as designed, energy is certainly already being directed to perform the work of chip manufacture—large amounts of energy, along with large enough amounts of information to manufacture the intricate layout of a very large-scale integrated (VLSI) circuit.

Investments in control, in particular, are investments in information that require resources, such as energy. Ever finer control requires ever more elaborate and detailed systems for sensing, for feedback, for processing, and for "executive" control. Manufacturing and operating these systems requires ever greater amounts of information and of energy to provide and process this information. Here, we see the two forms of Laughlin's mantras: On the one hand, the manufacturing process uses information to direct energy into work. On the other hand, the process of increasing the capability to use more information to more precisely direct work itself uses energy. Limits on energy and on information will, therefore, ultimately create new limits on approaching the efficient frontier.

Thus, we can see in applying the principle of energy–information trade-offs to a chip foundry how information can direct energy to do work and how information requires energy in order to do work, as well as how limits on either energy or information result in limits on the capabilities of a system to perform work.

Before departing from the foundry for the hospital, we might add one more perspective that could limit the performance of a system—energetic choke points, such as those described by De La Rocha, that are built into a system in ways that cannot easily be engineered away, as demonstrated by the choke point of Rubisco. More generally, in a living system, the path of evolution may be such that a choke point cannot readily be removed. In human-engineered systems, choke points can be more readily removed by the same kind of investment of information and energy that we have just described. However, the same limitations exist as for improving manufactured systems, described above and in Lightner's chapter.

To combine the perspectives of De La Rocha, Laughlin, and Lightner, the expenditure of energy and information to remove any given choke point may not be worthwhile from a cost–benefit perspective or may not be logistically feasible. For example, major rail lines to London all terminate around London instead of passing through. Although it is possible to draw lines on a map removing these choke points, the cost of building several major railroads through the center of London simply may not be worthwhile or feasible. Even when an alternative was eventually developed, of building one linking rail route, the cost of building a fast-enough link in very tight and expensive terrain impeded progress. (Likewise, major rail routes in the United States pass through Chicago for historical reasons, many of which no longer exist.)

To broaden our consideration of the choke point effect, what are some ways in which choke points get into human-designed systems, and what are their potential effects? Other chapters in the book remind us that these places where we want to improve performance are also places of employment—places that humans are engaged in work. We started the book by emphasizing that a richer understanding of work in living systems will remind us that human workers in a postindustrial workplace are not merely disembodied brains. Now we need to be reminded of that again by revisiting issues described earlier in the book that expand our perspective on human work.

The performance of a plant like a chip foundry is one of the most highly regulated and automated in existence. Most places of human employment will be less so. We have been reminded by Laland and Brown that humans have developed vast capacities for behavioral flexibility or plasticity and by Levin, Laland, and Saturay that this behavioral flexibility can easily override natural physiological balances. Thus, an optimal design of a place of employment—whether an iron foundry, a call center, or a neurosurgery wing—is essential. Performance–yield trade-offs and energy–information trade-offs exist just as much in a new neurosurgery wing as in a new chip foundry.

Let us start with the design of the neurosurgery wing itself. Blackwell has called our attention to factors in the design of these places that are difficult to override. The first of these is that designers—of chip foundries and of neurosurgery wings—are human. We are thinking not so much of human fallibility (or parameter variation) as of the combined effects of individual motivation and behavioral plasticity. Perhaps, as Blackwell has noted can be a frequent occurrence, the designers of either facility are strongly interested in winning the award for innovative design in their field. We are not just thinking of Blackwell's example of a design that is aesthetically elegant but nonfunctional. A designer might also develop a design that will unwittingly come far closer to the efficient frontier than the existing technologies and processes warrant in terms of parameter variation. The result would be that the neurosurgery wing will have a lower yield rate than necessary, not a result one would like when surgery is being performed on humans' brains. This possibility is all the more likely if the designer is unaware of where the efficient frontier is located, or even that there are such things as efficient frontiers or parameter variations (except as trial and error may have suggested, as in transoceanic flight). Today, therefore, the need for an integrative approach is pressing. One motivation for an integrative approach to the study of work is that there is much to be gained, for example, by having a designer who designs neurosurgery wings in a way that is informed by the knowledge of efficient frontiers and their consequences.

Blackwell also reminds us of a second human limitation. The designers of the neurosurgery wing are constantly engaged in attention investment decisions. These are their decisions about how to allocate their skills and energy to activities that they are undertaking. One might think that designers are well equipped to do this because they handle the effects of performance–yield trade-offs as they design. Indeed, the designers know that they have to pay attention to investment decisions at every moment. They encounter a myriad of familiar problems. Does the team designing the neurosurgery wing attend to patient comfort, physician performance, allied health personnel stamina, personnel safety, or the aesthetic preferences of the donor who is making the wing possible? Or all of these? The designers themselves are subject to attentional and performance trade-offs of the kinds described by Blackwell and Lightner. These trade-offs resemble the factors they should be taking into account as they design the wing, such as the energy–information trade-offs of the neurosurgery team members and the working-energy/take-home energy trade-offs of all of the parties who will work together in the wing. Can the design team actually attend to all of these?

The resources available for design and its execution are themselves limited. The constraint of the amount of the donor's money, which must go to design as well as to the facility itself, paired with the constraint of time, indicate that there are only so many designers who can work on the project over the design phase and only so much space that can be outfitted according to these various requirements. Given that resources are limited, attention investment allocations are being made throughout the design process. If attention is invested unwittingly, then these design factors will not all be attended to optimally.

Let us move on. The designers have finished, and the wing is up and running. Assume that the neurosurgery wing has, in the end, been designed to optimize performance by the neurosurgeon. In turn, that makes it more likely that the design is suboptimal in relation to the objective of the stamina of the nursing and other allied health staff. From this fully understandable allocation of resources and attention, there are any number of routes that now develop to suboptimal performance in the operating suite. We will trace only a few of them.

We mentioned specifically that a design consideration could have been the stamina of allied health personnel during long operations. That has been suboptimized. First, of course, that could directly affect allied health staff performance. Second, as Rosse and Saturay note in their chapter, these working conditions could trigger dissatisfaction and a variety of psychologically adaptive behaviors, including some that are counterproductive. Retaliation and neglect are not generally thought of as desirable during a neurosurgery, nor are these unheard of. Third, as Rosse and Saturay also note, via the Whitehall studies, the working stress on the nonphysician staff, coupled with their presence in a rigid hierarchy, has been shown in other settings to correlate with long-term impairment of health. In the shorter term and inside the operating suite itself, dissatisfaction of the allied health staff can and often does provoke (or result from) dissatisfaction in the physician, the neurosurgeon. Problems with staff interactions in general and increased requirements for physicians to work in teams in particular are two of many reasons frequently provided for impaired physician performance.

Nor are we yet finished with the possible impacts on the allied health staff. Levin, Laland, and Saturay hypothesize another route by which dissatisfaction can occur. Increased complexity of surgeries and increased surgery length, together with the suite's suboptimal design for staff stamina, can result in individuals being pushed beyond their internally regulated energy balance, presented as the balance between working energy and take-home energy. This performance envelope is largely invisible, as it occurs far from the absolute limits of human endurance. Levin, Laland, and Saturay hypothesize that, for example, pushing staff members past the limits of the working energy/take-home energy trade-off can trigger dissatisfaction on its own, with the potential results we have already described. Staff members pushed past these limits might also experience increased work-life conflict at home, with higher rates of turnover as one result, with the potential result in turn for a less qualified surgical staff.

The working environment we have portrayed brings us as well to the research described by Maier and Levin on the effect of responses to environmental challenges, including stressors, each of which can cause impairments to performance in our operating suite. First, working in the midst of an hours-long neurosurgery and being subject to direct hostility from the presiding neurosurgeon is an example of exposure to an inescapable stressor, just as it is for the neurosurgeon to be subject to having to conduct the delicate surgery in the midst of a hostile staff. The effects of responses to uncontrollable stress on energy mobilization and the capacity to work have been described by Maier and Levin.

Second, exposure to infection is not unknown in a hospital environment, and workers exposed to infectious agents will encounter sickness behavior and decreased capabilities for work, no matter their position or rank. Third, encountering hostile threats of the kind just described (e.g., "That'll be your job!" or "I'll report you to the Board!") are an example of exposure to an acute stressor, which we have seen can result in neuroimmunological responses that impair performance, even in the absence of an infectious agent.

All of these effects lead to the ongoing development of what Maier and Levin refer to as *intraindividual differences*—differences between one individual's performance from day to day and from week to week. A large hospital or university can be seen as rippling—or heaving—with the waves of these factors aggregated across individuals. Our smaller neurosurgery team is no less immune to such heaving from the intraindividual differences in performance of its members—as one knows from watching whitecaps form on a small mill pond or a beaver pond on a blustery day.

Those two ponds bring us back to a fundamental principle of understanding work in a broader context. Both the mill pond and the beaver pond are products of the process of niche construction presented by Laland and Brown, niche construction occurring at a simpler level than in the hospital wing or the chip foundry. When the mill was constructed, energy was invested to change the human niche and the natural niche. The need for the mill at the site of the mill pond has likely long since passed; today the pond is a remnant of the niche construction activity that continues to affect human and nonhuman life and work—for better or for ill. If one substitutes a steel mill, the same is the case, and the scale larger. As has been pointed out above, a natural byproduct of work (and of niche construction) are work's remnants, which themselves affect niches. In this section, we described the effects if architects are unaware of parameter variation and if neurosurgeons and hospital administrators are unaware of the effects of energetic trade-offs. In a similar way, niche construction theory calls us to another awareness—that the work we do today is constantly shaping, for better or ill, both the human niche and the niches of other living systems. Awareness of this dimension and effect of our work can only bring us benefits, as human workers and as coworkers with other living systems on Earth.

Moving Forward

Where does this leave us with regard to an integrative study of work in living systems? We started this volume with the study of how work gets done in a variety of non-human and even nonliving systems. We gained an understanding of the roles that various trade-offs and performance envelopes play in these systems and, of necessity, likewise play in human systems. We also explored aspects of human work. We learned of the continuity between work in other living systems and work in humans—both the similarities and the ways in which studying human work can give us access to knowledge that would be inaccessible in other systems. We studied examples of ways in which research in human and nonhuman systems, utilizing the continuity of work across living systems, can be brought together.

In Reflections, we have explored themes that were unexpected when we began our journey, observations made during the journey of patterns of work, and visual representations of both transformations and trade-offs that we can now, literally, see are part and parcel of how work gets done in living systems. We see better how work itself integrates life across levels of organization and across systems.

We have also taken those principles and applied them to examples to see how they can be combined for a richer understanding of work and performance. We have likewise looked at the limits of performance, in the chip factory and in the neurosurgery wing—inherent limits of how work gets done. Thus, our outlook on what can be gained in the study of living systems and in the world of work by studying work in this manner is one of optimistic realism, but not of utopianism, an approach that historically can too easily become bound up with any approach to understanding and especially to improving work.

Nor are we dystopian, another historically common accompaniment to the study of work, though our treatment in this section of the inherent limits on human performance based on various factors described in this book may strike some readers as pessimistic (or at least as moody). Instead, we believe that the relevant chapters do offer an important way forward in these areas: The performance envelopes inherent in work in living systems and further created in postindustrial work are the more difficult because they are at present largely invisible. We feel that increased understanding of these performance envelopes, of their widespread nature beyond human systems, and of their continued existence in postindustrial work can only help move toward better awareness of these factors and ways to work in harmony with them.

One example of bringing these approaches together in an optimistic manner— and one that suggests three key points of how the integration of the study of work moves us forward—was brought home to us recently by one of our colleagues who has become familiar with the approach of this volume and has, like others, suggested that one possible outcome of the publication of this volume is to build a community of people whose interests are compatible with the research and approach included herein.

While discussing a prospective project dealing with the application of technology to solving workplace and environmental problems, the conversation turned to the question of whether applying technology alone, without a more complete understanding of how work gets done, could successfully solve such problems on a robust basis. It soon became clear that, by sharing in common the approach of this volume, the conversation could move back and forth between topics related to this question that might be elsewhere considered to be in separated domains of natural and life sciences, social and psychological sciences, engineering, and practical issues of employment.

Had it been necessary to stop and clear passport control at each domain boundary, our conversation about the different issues would have been impaired. Moreover, at each of those boundaries, one of us would have been labeled a native and one an alien. At times, the alien would have been deemed to have lacked sufficient papers to cross, and, regardless, each crossing would have necessitated a pause, an interruption, and a reordering of a hierarchy between the two of us. At the same time, we would likely have also made no progress had the two of us simply been freely holding forth on such a range of topics without having a common framework about exploring work, a framework which could both shape our discussion and provide rigor.

The outcomes of conversations such as this represent the promise and the challenge of an integrative study of work in living systems. We believe that moving forward from this volume, the challenge and the opportunity will be to integrate this approach in three ways:

1. To integrate an understanding of how work in living systems gets done with research in other domains concerned with living systems.

2. To integrate a better developed knowledge of how work gets done in living systems with knowledge and challenges of human work and employment. With regard to this point, we strongly emphasize the term *integrate with* rather than *apply to*.

3. To integrate the knowledge gained in these first two activities with the fundamental challenge of integrating human work with the work of other living systems in a continuingly viable community of life, and work, on Earth.

Contributors

Alan Blackwell   Alan Blackwell is Reader in Interdisciplinary Design in the Computer Laboratory at the University of Cambridge and a Fellow of Darwin College, Cambridge.

Gillian Brown   Gillian Brown is Lecturer and Wellcome Trust Career Development Fellow in the Department of Psychology at the University of St. Andrews and Codirector of the Institute of Behavioural and Neural Sciences.

Christina De La Rocha   Christina De La Rocha is Professor in the Marine Environmental Sciences Laboratory (LEMAR) of the European Institute for Marine Research (IUEM) at the Université de Bretagne Occidentale, Brest, France.

Kevin Laland   Kevin Laland is Professor in the School of Biology at the University of St. Andrews and a Fellow of the Royal Society of Edinburgh.

Simon Laughlin   Simon Laughlin is Professor of Neurobiology in the Department of Zoology at the University of Cambridge; a Fellow of Churchill College, Cambridge; and a Fellow of the Royal Society.

Robert Levin   Robert Levin is Director and Fellow of the Center for Integrative Study of Work (CISW) at the University of Colorado at Boulder and Scholar-in-Residence in the university's Graduate School.

Michael Lightner   Michael Lightner is Professor and Chair in the Department of Electrical, Computer, and Energy Engineering at the University of Colorado at Boulder and a Fellow and the 2006 President of the IEEE (Institute of Electrical and Electronics Engineers).

Steven Maier   Steven Maier is Distinguished Professor in the Department of Psychology and Neuroscience and Director of the Center for Neuroscience at the University of Colorado at Boulder and a Fellow of the American Association for the Advancement of Science, the American Psychological Association, and the American Psychological Society.

Joseph Rosse Joseph Rosse is Professor of Industrial and Organizational Psychology in the Division of Management and Entrepreneurship, Leeds School of Business, University of Colorado at Boulder.

Stacy Saturay Stacy Saturay is Instructor in the Division of Management and Entrepreneurship, Leeds School of Business, University of Colorado at Boulder.

Index